LIVING WITH OUR GENES

ALSO BY DEAN HAMER AND PETER COPELAND

The Science of Desire

LIVING WITH OUR GENES

WHY THEY MATTER MORE THAN YOU THINK

Dean Hamer and

Peter Copeland

DOUBLEDAY

NEW YORK LONDON TORONTO SYDNEY AUCKLAND

PUBLISHED BY DOUBLEDAY
a division of Bantam Doubleday Dell Publishing Group, Inc.
1540 Broadway, New York, New York 10036

DOUBLEDAY and the portrayal of an anchor with a dolphin
are trademarks of Doubleday, a division of Bantam
Doubleday Dell Publishing Group, Inc.

Book design by Ellen Cipriano

Library of Congress Cataloging-in-Publication Data
Hamer, Dean H.
Living with our genes : why they matter more than you
think / Dean Hamer and Peter Copeland. — 1st ed.
        p.   cm.
1. Temperament.   2. Temperament—Physiological aspects.
3. Personality—Genetic aspects.   4. Behavior genetics.
        I. Copeland, Peter.   II. Title.
            BF798.H35   1998
155.2′34—dc21   97-29818
                                                    CIP

ISBN 0-385-48583-2

5 7 9 10 8 6

To our children
Addie, Isabella, and Lucas

Nature is often hidden, sometimes overcome,
seldom extinguished.

—SIR FRANCIS BACON

CONTENTS

EMOTIONAL INSTINCT

The Genetic Roots of Personality

We are all sensitive people.
—MARVIN GAYE

I am what I am.
—POPEYE

The invitation to her twenty-fifth high school reunion came out of the blue, and Janice was surprised anyone had been able to keep up with all her address changes over the years. Janice, 43, graduated from a large high school in the Midwest. Her parents were not wealthy, so throughout school she worked part time to afford the clothes and accessories she deemed essential. That didn't leave much time for homework, but she managed to get good grades. Janice had many girlfriends in high school and even more boyfriends, but after graduation she never bothered to keep up and had never attended a class reunion. She just didn't see the point; that was then, and she had grown way beyond those days.

When she received the invitation, her first thought was: boring.

After high school, Janice had married an older businessman. He wasn't much of a lover, but he put her through college, taught her about the world of finance, and broadened her horizons. Eventually the relationship faded and there was a divorce; he was a sweet guy, but he just wasn't that interesting anymore. After the divorce, Janice moved to Southern California and used a sizable settlement to jump into the real estate business. There were many ups and downs in the business, but during the last recession she gambled on some large properties and was doing very well indeed. Recently she purchased a white BMW convertible with a tan leather interior.

Janice went through a string of lovers after her divorce but never remarried. She didn't want to be burdened with kids or waste time keeping house. Recently, she started a passionate affair with her yoga instructor. Janice sends him flowers (usually a single orchid or some other exotic) at the ashram and has taken him on a vacation to Sri Lanka, where they made love on a palm leaf mat on the beach. She feels like she is making up for lost time, romantically and spiritually.

Proud of her striking figure, Janice has worked hard to stay at the same weight since graduating. For many years she was addicted to diet pills, but recently she joined a self-help group and now is drug and alcohol free. Smoking has been harder to quit because of her fear of weight gain. The invitation to the reunion got her to thinking about how far she'd come since high school. The next day she decided, "What the hell. It'll be a hoot." She made plans to fly out in the afternoon and return early the next morning.

Ralph is also 43. In high school, he was serious about academics and maintained a steady B average. He was never part of the "in crowd," nor was he a jock, but he's kept up with a few close friends and has never missed a class reunion. His yearbooks are arranged on a shelf by his bed. Ralph went to the state university and obtained a degree in environmental science. He had wanted to attend a private school in the East, but his grades weren't good enough to get a scholarship (for which he blamed his teachers' lack of interest), and his family couldn't afford the tuition (for which he blamed his parents' lack of ambition).

Ralph's first job was with the state wildlife commission, and he's been there ever since, rising from research assistant to middle management. The money is lousy, but he likes the work, and he loves nature. His favorite activity is walking along the local lakeshore, collecting seashells and polished glass. He despises the proliferation of high-rise developments along the lake and has written many angry letters to the editor of the local newspaper. Sometimes he gets so upset that he loses sleep and can't focus at work.

Ralph has been married for 25 years and has three children. He and his wife have a harmonious albeit no longer passionate relationship. Ralph chose their suburban house because it was close to the bus line. He doesn't like to drive, and his wife chauffeurs the kids in a Ford station wagon. Over the years, Ralph has lost most of his hair and has gained more than a few pounds, mostly around the middle. He's never smoked and drinks rarely.

When Ralph got the invitation to the twenty-fifth reunion, he confirmed his place by fax the same day. "It will be good to see how everyone's doing," he thought, truly relishing the memories of the good old days.

On the night of the reunion, Janice is looking right at Ralph but doesn't really see him. It takes her a while to recognize him—he wasn't the type of boy she dated—but eventually she remembers him as the guy who was always standing at the edge of the crowd, just like he is now. There was a time when Ralph had a wild crush on her, but Janice never paid much attention. "I sure hope I don't get stuck next to *him* at dinner," she thinks. "What a nerd!"

Ralph spotted Janice, too, and his stomach tightened nervously. Janice was once the object of his most passionate fantasies, regrettably all unrequited. Her hair color and the shape of her nose had changed, but not her confident way of carrying herself. Confident was a nice way of putting it. Stuck-up would be more accurate. "Oh boy," thinks Ralph. "Hope I don't end up next to her at dinner."

Of course they end up seated together.

Ralph says, "Janice, it's really great to see you!"

Janice replies, "Why Ralph. What a pleasure! It's been ages!"

Wine is served. Janice refuses, but between courses she sneaks to the bathroom for a smoke. Ralph drinks half a glass of Chablis and feels himself start to relax. He eats another roll and passes the basket to Janice, who passes them on without taking one. She whispers to Ralph, "Look at Nancy Abramms will you. I can't believe how fat she is. She used to be totally into how petite she was! Not anymore."

"Well, you know what happened to her," Ralph confides, leaning into Janice's ear and smelling her perfume. "Divorced. Husband left her for another woman. Very messy."

Janice picks at her plate, and Ralph, eager to keep up the conversation, asks about her marriage.

"Forget marriage," Janice says with a dismissive wave. "I'm into romance now."

Ralph feels his face glowing red, but he's also a little aroused. He takes a big drink of wine. "Romance?"

"I'm seeing this fantastic guy. He teaches yoga at the club where I go, and when we met it was like we had known each other before or something. It was this incredible bond."

Ralph attempts to tuck a roll of flab into his pants. It's too uncomfortable, so he blouses out his shirt to soften his waistline. He sits back in the chair, ready for the after-dinner speakers. He wonders what's for dessert. He searches for something clever to say about how Nancy Abramms doesn't need any dessert, but then Janice stands in one fluid motion, drops her napkin on her half-filled plate and starts to leave. Ralph looks up at her until Janice bends to give him a good-bye kiss, lingering just a little longer than a peck, long enough to make the back of his neck tingle like he can't remember when.

"Same old Ralph," she says, meaning it in a nice way. "You're just the same."

"I guess I am," Ralph says, resigned to the truth of it. "You make me feel like I'm right back in high school. You haven't changed one bit."

What makes Ralph "just the same" as he was 25 years ago, despite having added 40 pounds, a wife, three kids, and a career? How can he think that Janice hasn't changed when she has changed so much, from her California makeover to her economic and social status? What Ralph and Janice recognized in each other was not physical appearance, possessions, or activities. It was their core personalities, much of it hardwired into their bodies since birth, a genetic legacy from their parents as surely as the color of their eyes.

The dictionary definition of personality is, "The sum total of the mental, emotional, social, and physical characteristics of an individual." It's personality that determines the way you react to others, the way you communicate, the way you think and express emotions. These are the outward manifestations of basic traits that characterize a personality throughout life. Your thoughts, fears, hopes, reactions, behaviors, and dreams all come from this core personality.

Personality determines not just being but behaving. It influences how much you eat, drink, smoke, and sleep. Personality determines whether you are aggressive or shy, active or passive, who you are attracted to, and what you want to do with them if they'll let you. It influences the amount of stress in your life, your physical health, and whether you live in pain or pleasure, in a sleepy haze or a high-octane blur. Personality is so complex that even though millions of people have walked the Earth, no two have ever been the same. Just as the physical body has a seemingly unlimited variety, so does the personality that makes the body get up and go. Personality is what makes each person unique.

The latest research in genetics, molecular biology, and neuroscience shows that many core personality traits are inherited at birth, and that many of the differences between individual personality styles are the result of differences in genes. When you are conceived by two people, you are created from their genes. You are the product of generations of evolution, countless bits of information collected over millions of years, focused, narrowed, and refined until you were pushed out of the birth canal into the world. You look like the people in your family—and in some respects you feel and act like them, too. You have about as much choice in some aspects of your personality as you do in the shape of your nose or the size of your feet. Psychologists call this biological, inborn dimension of personality "temperament."

Just because a person is born with a particular temperament, however, doesn't mean there is a simple set of instructions or blueprints. Nor does temperament mean that people are "stuck" with their personalities from birth. On the contrary, one of the marvelous features of temperament is a built-in flexibility that allows us to adapt to life's hurdles and challenges. Growing up means not only learning the ways of the world, but also how to deal with yourself. Psychologists call this more flexible aspect of personality "character."

Everyone has the ability to grow and to change at every stage in life. People can learn from experience, from parents, and from friends. Individuals have the option to give in to temperamental weaknesses, or to overcome them. They can take advantage of temperamental gifts or hide them. People can indulge their desires to smoke, drink, or eat too much—or they can resist. Sometimes they will do both during the course of their lives.

It was Ralph's cautious temperament at work when he saw Janice and had the gut reaction, "Danger!" But when he greeted her cheerily, it was character making him polite. When Janice spied Ralph and felt an almost physical revulsion, it too was her temperament. But when she made an effort at conversation, her character was kicking in. Both have learned to push aside their "gut" instincts to get along with others.

Together, temperament and character form personality. Temperament and character come from different places in the brain and are expressed in different ways. By digging deep into personality, past the clutter of myths, learned responses, and stereotypes, anyone can discover his or her underlying temperament and develop positive character traits to become the kind of person he or she wants to be.

Temperament is the focus of this book: what it is, how to recognize it, and where it comes from. Psychologists usually use "temperament" to refer to attitude or affect (the way we see the world), but we use the broader definition to include all characteristics including behaviors (what we do). Many people want to deny that inborn temperament exists, preferring the myth that we are born blank slates and are purely products of our environment. We want to believe that we can remake ourselves into anything we want, even when our attempts repeatedly fail. But no matter how hard he tries, Ralph will never be as outgoing as Janice, and Janice will never be as loyal as Ralph. They both are happy in their own way, but they will never be alike. The same is true for everyone: there are traits you can change and others you can only attempt to control or modify. You cannot be anything you want to be, but as they say in the army, it is possible to be all you can be.

The "environment" matters, of course, but contrary to popular belief, the most important environmental factors are not rearing, education, or social status. Rather they are random and uncontrollable experiences such as the precise concentration of a particular chemical in the brain, or something apparently minor like a childhood case of the measles. While we like to imagine ourselves to be the carefully crafted products of our upbringing and education, we are actually shaped by the same sort of chaotic events that make each snowflake unique.

Just as a snowflake has little to say about its shape, we are born with limits on the shape of our bodies, the color of our skin, and type of hair. We spend billions of dollars—and hours of sweaty agony—trying to mold our bodies into the current cultural ideal, only to watch them sag back to their natural shapes. Try as we might to change our shape, most of us will fail. We will eat the way we've always eaten, and we'll have the same activity level we had as children, or even earlier in the womb. The reason, scientists have learned, is that body weight is more determined by inheritance than by any other factor. Experiments have shown that mice with a certain type of gene grow fat even when they are given almost no food. Humans contain an obese gene that is almost identical to the mouse version, and some people will have a harder time controlling their weight, not because they are weak or eat too much, but because the genetic bar is set higher. How quickly the body breaks down with age also is controlled by genes, and scientists recently have been able to genetically engineer simple animals with life spans double or even five times longer than normal.

A similar controlling role is played by the genetics of gender—the biological differences between men and women. Every fetus is created without a sex until a single gene switches on and begins a cascade of chemical reactions that turns half of us into males and half into females. The changes affect not only physical characteristics but mental ones as well. Men are programmed to seek more partners and sexual novelty; women are "serial monogamists," seeking mates who will remain long enough to raise offspring. Women want emotional attachment and financial security not because that's what they are taught but because it helps the species survive.

Other behaviors determined largely by heredity include addictions to alcohol, tobacco, and dangerous drugs. Scientists now believe that it doesn't matter what substance you abuse, but why and how you got hooked. Understanding that addiction is a physical condition, a condition that changes the actual functioning of the brain, is vital to recovery. Violence and aggression also have genetic roots. Some people are born with shorter fuses and are more likely to lash out at others. Numerous studies show that altering the levels of a single brain chemical can completely change an animal's level of aggression. A simple manipulation of a single gene can turn a meek and mild rat into a crazed killer. The same brain chemicals are found in humans, and certain people are "driven" to violence by a force within them. Their lifelong struggle will be to consciously override what they are programmed to do.

How we think is also a product of genes. The evidence that IQ is largely inherited is overwhelming. Some genes determine how quickly the brain can process information. Others may control particular circuits, such as those for mathematical calculation or perfect pitch. What we've always called "God given talents" are known in laboratories as genetically endowed traits. The encouraging news is that genes don't always play their strongest role until adulthood. Intelligence in children can be very strongly influenced by adults because infants and youngsters are not capable of stimulating themselves intellectually; they have to be taught and exposed to new things. The innate ability to learn more than one language, for example, may be shared by many people, but only those given a chance early will become bilingual. Generations of frustrated people who didn't begin studying a foreign language until high school are proof of how much more difficult it is for the mature brain to learn.

My own lab at the National Cancer Institute made headlines around the world when we discovered a genetic link to male homosexuality. That research has been expanded to look at sexual orientation in women, and the surprising results are discussed here. Since the discovery of the "gay gene," my lab has gone on to find genes for two other personality traits: novelty seeking and worry. Novelty seeking means the desire to seek out new experiences or thrills, and our work shows that this is largely an inherited predisposition. The other gene is for harm avoidance, which makes people anxious, fearful, and shy. Fascinating studies show that shyness, or being outgoing, is inherited at birth and lasts a lifetime. My lab discovered that harm avoidance is influenced by a "genetic Prozac," a natural mechanism inside the brain that controls the level of anxiety and has great potential to ease depression.

The emerging science of molecular biology has made startling discoveries that show beyond a doubt that genes are the single most important factor that distinguishes one person from another. We come in large part ready-made from the factory. We accept that we *look* like our parents and other blood relatives; we have a harder time with the idea that we also *act* like them. In other species, we value and encourage genetic differences in "personality." Consider the difference between a Wisconsin dairy cow and a bull from Pamplona, or a golden retriever and a pit bull. Human breeding is less orderly, but children do share personality traits with their parents. Every grown man has experienced a shock of realization when he does something exactly like his father before him. Every mother has a similar experience when a child behaves exactly like her. This is not bad; it's beautiful. This does not mean we are doomed to become our parents; it means we begin our journeys where our parents left off.

Those who accomplish the most—measured in money, intelligence, skill, happiness, or love—are the ones who make the most of their genetic inheritance. If identical twins, who share exactly the same genes, can turn out differently, that means that genes are not fixed instructions. More than a musical score, genes are like the musical instruments. Genes don't determine exactly what music is played—or how well—but they do determine the range of what is possible. Imagine that every person is born a seed or an acorn; all their potential is compressed in that tiny form. Whether they reach their potential as mighty oaks depends on many factors, but they are born as unique individuals with their own distinguishing characteristics.

TEMPERAMENT: FROM CRADLE TO GRAVE

· Were you an active child who bounced off the walls, opening closed doors, and pulling things out of drawers? Or were you content to sit in mother's lap, twirling her hair in your fingers, and watching the world go by?

· Were you the kind of child who was upset by change?
Did the arrival of a new babysitter cause screaming and
footstamping, or curiosity and a desire to show off a pretty
dress? Did a novel situation feel like a threat or an
adventure?

· Did you find yourself happy one minute and sad the next?
Did some days seem far better than others for no real
reason? Did your moods swing back and forth, or were you
generally calm and on an even keel?

These questions describe three measurable aspects of
temperament: activity level, reactivity, and mood. As the examples suggest, these characteristics are expressed early in
life. They are not learned from parents or books, nor can
they be easily controlled through will power. A baby doesn't
decide that a new face is scary, it just is. A small child is not
active and inquisitive because it wants to be, it's born that
way. A toddler doesn't feel truly and inexplicably sad just
because she fails to get her mother's attention. More than
learned responses, these are constitutional ways of being, as
much biology as psychology. Temperament exists at the level
of instinct, which means a natural and inherent aptitude, impulse, or capacity.

Temperament is not easy to change; it tends to endure as
a person matures. If you were a shy baby, you probably are a
shy adult. If you were an adventurous child, you still probably enjoy doing new things. If you were a sad child, most
likely there still are days when you don't want to get out of
bed.

Individual differences in temperament are produced in
part by biology, just like the shape of a nose or the color of

skin. The instructions for human development, including aspects of temperament, are carried in genes passed from parents to children. Although there are important nongenetic factors, such as parenting style and schooling, no single influence is more profound than genetic makeup. That's why temperamental traits expressed early persist throughout life: we are born with the same genes we die with. But the genes themselves don't make a baby cry or giggle, or make the difference between a gregarious car salesman and a shy data processor. Rather, the genes control certain aspects of brain chemistry, which in turn influence how we perceive the world and react to that information.

Temperament does not come fully formed with a new baby. Instead, the baby is born with the potential to acquire a temperament in response to the environment. Not only do genes predispose people to certain ways of being, genes also appear to have a role in what experiences we seek, pushing us toward certain environments that will shape our behavior. Obviously a baby is not born with the full range of human emotion completely developed; only the fortunate among us will achieve that during an entire lifetime. So temperament *is* learned, but not in the way we memorize a telephone number. Rather people "learn" temperament through *emotional* memory, what most people know as habit.

If a baby is frightened by a new face, chemical reactions in the brain make the baby feel anxious and afraid. Attentive parents can try to make seeing a new face a fun experience, but the brain's reaction is stored in the baby's emotional memory. This is not a one-time occurrence, but a pattern that reinforces itself. One scary Halloween mask is not going to permanently scar a child, but rather many reactions build the emotional pathways in the brain. Later in life, it would not be surprising if the extremely shy baby grows into a person anxious about meeting new people. In the same way, if a baby learns that climbing onto the couch, peering out the window, and waving cheerily at the faces passing by makes the brain feel good, that feeling also is stored in emotional memory. The outgoing baby could grow into a "people person" who enjoys making new friends.

But why are some babies shy and others outgoing? The shy baby and the outgoing baby are responding to the same new faces, but why does the first baby's brain react negatively to a stranger, while the second baby reacts positively? The root of these responses is in the genetically determined chemistry of the brain, especially the primitive part of the brain called the limbic system. The limbic system is responsible for emotional behavior—the way people feel—by generating "gut reactions," the most powerful drives, behaviors, and feelings, often the ones that feel beyond the control of consciousness. Deep in the limbic system are the roots of fear, aggression, lust, and pleasure.

If everybody had the same genes that built the same limbic system, and then had the same life experiences, all personalities would be the same. But limbic systems are different because genes are different. Experiences are different because we live in a world with so many possibilities. No two people, even identical twins raised in the same home, can share the *exact* same experiences, which is part of the reason for the unlimited variety in the second dimension of personality—character.

CHARACTER: LIVE AND LEARN

· Are you willing to endure personal sacrifices to make the world a better place? Or do you think it's each person for himself?

· Do you usually accept people as they are, even if very different from you? Or do you want everyone to act the same as you?

These are aspects of character. People aren't born with these beliefs, they acquire them—from parents, friends, teachers, and spiritual leaders; from doing things right and remembering how they felt, or doing them wrong and being punished.

The memories that form character are mediated by the cerebral cortex, which remembers people, places, and things and allows us to calculate, compare, judge, and plan. The reason that character is the most distinctly human aspect of personality is that the cerebral cortex underwent a dramatic burst in size and complexity in recent evolutionary history and is much larger and more advanced in humans than in primates and lower ancestors. The cortex is the manager for the rest of the brain, analyzing the world and deciding how to respond.

Although temperament and character might seem to be independent parts of personality, they are intertwined. The wonderful thing about character is its ability to modify temperament, to allow people to take advantage of the useful parts of temperament and downplay the less desirable biological tendencies or instincts.

At the heart of character is the concept of *self*. Do we

see ourselves as responsible for our own actions or at the mercy of outside forces? Are we an integral part of society like Ralph or a "free agent" like Janice? Of all the things we learn and remember in life, the most important is who we are.

Although the initial responses to stimuli are determined by the largely inherited temperament, the way people interpret and act on those responses depends on the acquired character. In the case of a woman we'll call Alexandra, going to a party was agony. She was a shy child who grew into a shy adult, but her job required frequent social engagements. In order to advance, she had to overcome her natural fear of social situations. At first, the effort was extremely stressful, but she found that the more she pushed herself, the easier it became. Doing something "against her nature" was rewarded with praise from her boss and actually started to become fun. In time Alexandra was able to reprogram her temperament so much that she came to regard social events with pleasant anticipation rather than dread.

What Are Genes For?

Alexandra was naturally shy largely because of her genes, but genes are not switches that say "shy" or "outgoing," or "happy" or "sad." Genes are simply chemicals that direct the combination of more chemicals. The chemical that makes up genes is called deoxyribonucleic acid, or DNA. It comprises simple building blocks called bases, of which there are just four different varieties: A, G, C, and T. The bases come together in long strings, and each DNA molecule consists of two of these strings paired according to the rule that A matches T and G matches C. DNA stores information in the order of bases. The DNA sequence "AGCT" means one thing, and the sequence "TCGA" something else, as different in meaning as "taps" and "spat."

The information in DNA is converted into proteins, which are made of amino acids. Proteins are where the action is. The most important function of proteins is to act as enzymes that change one chemical into another. For example, it's an enzyme that converts tyrosine, an amino acid found in many foods, into dopamine, a brain chemical that can make you feel active and excited. A different enzyme breaks down the dopamine into smaller molecules and thereby leaves you feeling more relaxed or even lazy. Different enzymes make and degrade the more than 300 brain chemicals that influence thinking, acting, and feeling.

The brain and body are built by DNA, and everyone's DNA is pretty much the same. We all have 99.9 percent the same DNA as Michael Jordan, Albert Einstein, Elizabeth Taylor, Charles Manson, Julius Caesar, Julia Child, and Jules Verne. All of them and everyone who has ever lived have the same 100,000 or so genes, which are organized into the same 23 chromosomes.

But "pretty much the same" is not exactly the same. There are differences in DNA—about 0.1 percent, or 1 bit out of every 1,000. Considering that there are 3 billion chemical bases in question, the differences matter. Where Michael Jordan's DNA says "G," Michael Jackson's may say "C," and Andrew Jackson's said "T," while Jack the Ripper's said "A." There are roughly 3 million such differences between individuals, and these differences are responsible for all the inherited aspects of the variations among people, from eye color to height to personality or intelligence. It's hard to believe that such a tiny difference—one-tenth of 1 percent—could make such a great difference in how people turn out, yet this percentage is actually an exaggeration. Many of the 3 million variations don't mean anything as far as we know, so there are even fewer differences that matter.

If you still don't believe that 0.1 percent of DNA could be responsible for so many differences, consider the fact that human DNA differs from chimpanzee DNA by only 1 to 2 percent; your DNA and a chimp's are at least 98 percent the same. Yet this seemingly "fine print" of DNA instructions is the reason humans can do calculus, compose poetry, and build cathedrals, while chimps pick bugs off each other and eat them. Humans have pretty much the same DNA as a chimp because that's where we came from; and the chimp is close to the ape because that's where he came from; and so on down the line to fish and reptiles and even single-cell organisms such as yeast. This evolutionary conservation has a beneficial side effect: we can often figure out what a human gene does by looking at a similar gene in simpler organisms.

One of the most common misconceptions about genetics is that there are genes "for" things. Some people have the genes "for" breast cancer, shyness, blue eyes, and so they must have the disease, condition, or trait. This is what people tend to think when they hear about a gene "for" depression, or a "gay gene," or "obesity gene." If that were true, it would be easy enough to undergo testing to see what genes you have, and therefore what you ought to worry about. That's not the way it works. Everyone has a "mood gene," and a "sexual orientation gene," and a gene that regulates body weight. The difference is that the genes come in different varieties or flavors.

For example, maybe the "mood gene," which everyone

has, makes a receptor protein that responds to the hormones released under stress. Maybe the only difference between two people is that one has a gene with T at position 4,356, whereas the other person has a C at the same spot. That might be enough to affect the strength of the electrical current flowing through the cells, so the same amount of hormone produces a gentle buzz in one person and a walloping jolt in another. That single detail—1 letter out of 3 billion—could mean the difference between a mostly cheerful person and one who is easily depressed. Both people have the same gene, but the fine print makes all the difference. Imagine a room filled with 30,000 books. Here the difference would be the equivalent of a single letter in a single book.

BEHAVIOR GENETICS

We know that DNA makes unique brains, and each brain develops in a unique environment to form an individual personality. The question is whether the differences in DNA make us different. Searching for the answer is the science of behavior genetics.

Behavior genetics is so new that the first professional journal in the field appeared just 27 years ago—a heartbeat in the history of science. Its roots, however, go back a century to Francis Galton (1822–1911), a British biologist who was a cousin of Charles Darwin. He was fascinated by genius, perhaps because he himself was a child prodigy who could read and write before he was three and grew into a remarkably talented adult.

Galton's famous experiment was to compare 35 twins who were "alike at birth" with 20 pairs who were "unlike at birth"—what we now recognize as identical twins and fraternal twins. He concluded that twins who were physically alike at birth grew up to be alike as adults—not just in their physical characteristics but in their personalities, intelligence, interests, and careers. Galton wrote, "There is no escape from the conclusion that nature prevails enormously over nurture when the differences of nurture do not exceed what is commonly to be found among persons of the same rank of society and in the same country. My fear is that my evidence may seem to prove too much, and be discredited on that account, as it appears contrary to all experience that nurture should go for so little."

Galton's work was discredited, but not for the reason he anticipated. He took his studies to what he considered the next logical step: improving the human race by breeding "good" stock like himself and weeding out the undesirables. Today he is often best remembered as the father of eugenics, a movement whose most infamous advocate was Adolf Hitler.

Galton was correct, however, about the significance of the two types of twins. Identical twins develop from a single fertilized egg and therefore are exact genetic replicates. Fraternal twins grow from separate eggs and are only as genetically similar as ordinary siblings. Thus by comparing the resemblances and differences between identical twins, fraternal twins, and unrelated individuals, it's possible to calculate how much of a difference in a trait is caused by inherited factors. This is known as heritability.

The drawback in studying most twins is that they usually grow up together—in the same household, with the same parents, in the same neighborhood, in the same country, at the same moment in history—so maybe they are alike because they learned the same things. The solution is to look at identical twins who were separated at birth or shortly after. The most famous such study is the Minnesota Study of Twins Reared Apart, in which scientists have tracked down people who didn't even know they were part of a pair. The resemblances between these twins can be truly eerie.

Jim Lewis and Jim Springer were identical twins who were separated at birth and reunited for the first time when they were 39 years old. At their meeting, they both were six feet tall and weighed 180 pounds and were so similar in appearance that strangers could rarely tell them apart. The physical resemblance was to be expected, but when the two men sat down to talk for the first time, they found some rather strange similarities. Both had been married twice, first to women named Linda and then to women named Betty. One had a son named James Alan, the other James Allen. As boys, they both named their dogs Toy. Thirty-nine years after they were separated, they both smoked Salem cigarettes and drank Miller Lite.

After meeting the Jim twins and studying hundreds of other such pairs, Thomas J. Bouchard, Jr., and his fellow researchers concluded that identical twins reared apart are nearly as much alike as twins raised together. This finding was so shocking when it was published in 1988 that some people refused to believe it. It didn't seem possible that two people who had never met could be as similar as two people raised as brothers or sisters in the same home. The evidence, however, was strong, and it showed that genes not only help determine how we look, but how we act, feel, and experience life. In case after case discovered by researchers, nature won out over nurture.[1]

Amy and Beth were born identical twins, cloned from

the same egg and sharing the same genes. At birth, they were given up for adoption to separate families, but they were tracked carefully for the rest of their lives. The new parents—in both cases Jewish families in New York State—were told that the little blond girls were part of an experiment, but not one that concerned twins. Amy's family was much better off economically than Beth's, but Amy's mother was overweight and insecure, especially about her role as a mother, and she felt jealous of the pretty new baby. Beth's mother, on the other hand, was bright, attractive, and full of love for her new treasure.

Amy never really fit in to her new family, and grew up with emotional problems. Even as an infant, she was cranky and needy, traits that changed only in how they were expressed as she developed from a difficult baby into a troubled adolescent. She "refused" to grow up, had difficulty learning, and suffered greatly as a rejected child.

Beth grew up in a loving family and was the center of attention from the moment she was brought home. Her parents doted on her and gave her all the love in the world. According to the theory that we are products of our environment, Beth should have turned out beautifully. Where her sister Amy might have had an excuse for her troubled personality, Beth had every advantage. But Beth fussed and cried as an infant, just like Amy. As a little girl, Beth was afraid of everything and fell behind other children. As an adolescent, she was impossible. Despite the apparent advantages of her home life, she turned out exactly the same as Amy, the sister she never knew. Like bullets fired from the same gun, the girls followed trajectories that were remarkably similar.

The story of Beth and Amy is unsettling because parents

want to think we have a role in how our children evolve, and we all want to believe we are the product—at least in part—of our own free will. The good news is that parents do have a role, and individuals do have a say in the type of people they become. Even Galton himself didn't believe he was a genius solely because of his blood. If he had been denied any contact with humans as a child, he likely would have been regarded as retarded or feral. If he had been brought up in a lower-class household, he might have ended up as an exceptionally clever butler—but he never would have been called "Sir." On the other hand, if Galton had been born with a less extraordinary set of genes, no kind of upbringing would have produced such an extraordinary scientist.

The advantage that we have today is that our understanding of genes is growing so quickly. Understanding genes does not mean resigning ourselves to some preprogrammed fate. This is a tool for liberation, a scientific window into the soul. Yes, we are born with a certain genetic makeup. No, that doesn't mean we have no control over our lives. No scientist, nor observant parent, really believes we are born as completely blank slates to be filled in by our upbringing. The key is the interplay between the hardware we are born with and the software we add. It's not nature *or* nurture, it's nature *and* nurture. In fact, it's part of our nature to respond to nurture.

For years, however, one part of the equation has dominated. Nurture was considered far more important than nature. The study of personality was led by psychiatrists and psychologists who focused on the impact of childhood experience and trauma. According to these experts, understanding environmental influences was the only thing necessary to understand a human being. It was as if all people were identical, except for the experiences that shaped them. What's more, the same experience was supposed to affect each person the same way. A loss, an abuse, a distant father, or a smothering mother produced predictable, quantifiable, and comparable results in all people.

This theory is not only stupid but cruel. People are different because they have different genes that created different brains that formed different personalities. The real breakthroughs in understanding personality are not occurring on leather couches but in laboratories. Some of these new findings from labs around the world are explained here for the first time. The lessons can be applied to your own life and to the lives of your children.

Understanding the genetic roots of personality will help you "find yourself" and relate better to others. The knowledge can help you in relationships and at work. It should also ease the burden a little on parents obsessed with providing the best environment for their children. Giving children love and knowledge is as essential as giving them food, but at some point, parents must understand that children are already on a path beyond anyone's choosing. Children are who they are, and parents are better off getting to know their own children than trying to mold them into some ideal created out of thin air. Children are to be discovered as well as shaped; they should be allowed and encouraged to develop to their own potential. People are unique from the moment of conception, they do not begin as indistinguishable lumps of stone sculpted by life into individuals. Each of us is born into the world as someone; we spend the rest of our lives trying to find out who.

ONE

THRILLS

Getting High on Life

Thrills and pills and daffodils will kill.
—THE ARTIST FORMERLY KNOWN
AS PRINCE

Charles O'Rourke walked into the room like he owned the place. He had sandy hair, a craggy chin, and dark wraparound sunglasses. His impeccably tailored suit hung easily from big shoulders. He moved like a leopard, never stopping, agile and alert. He filled the room with his presence, sending out a powerful message: I'm in charge here.

By day, Charles, 42, was a stockbroker. His clients were like him—young, rich, and planning to make more money, lots more. They were not patient and they didn't mind a little risk. In fact, it was the risk they liked. They called Charles every day looking for the big tip, the one piece of information they could parlay into a fortune. Charles worked the phones, calling middle managers of companies he followed to get on the inside track. Some days he couldn't bear to be in the office, so he'd drive out to a high-tech industrial park.

There were a couple of companies he was tracking, and he liked to check them out firsthand.

The thrill of the market was what attracted Charles, not so much the money. The money was nice, but he liked the chase. The ticker made his heart pound fast, and he rode its ups and downs like a jockey. He lived for excitement. His pace during the weekend was no different; only the setting changed. He had a red convertible, buffed to a loving shine and fitted with a rack to carry a windsurfer. At the beach, he would head straight into the waves, cutting through the water toward the horizon. He surfed until his arms ached and his legs trembled. He loved how the salt tingled his taut skin, red from the sun and glowing from exercise. At night, he partied with friends, getting recharged for the next day. He couldn't wait to get up on Monday morning to jump back into the market.

Charles had a younger brother who lived in the Northeast. Not a glamorous location, but a pleasant enough city. The brother, whom we'll call Michael O'Rourke, rented a little apartment that was inexpensively decorated but meticulously clean. His pride and joy was a window box filled with geraniums. He loved to sit alone at the kitchen table, sipping herbal tea and watching the little plants take the sun. Michael lived less than half an hour from where he and his brother had grown up. Charles had moved to the West Coast, but not Michael; there was no reason to go so far away. He knew his way around and had a few friends and relatives nearby; he felt comfortable there.

Michael, 28, looked like a young and somewhat slimmer version of Charles: the same rosy complexion and reddish hair, even the same easy smile. He dressed casually in button-down shirts and corduroy pants. He was in graduate school studying to become an elementary school teacher. He had been studying to be a microbiologist, but switched to teaching because he found science was too stressful. He didn't smoke and drank infrequently. He didn't windsurf but he liked to jog, alone and when the streets were deserted. His idea of a fun weekend was to visit the local museums, though he always waited for the week after an opening when the exhibit was likely to be less crowded.

Charles and Michael were born of the same mother and father and were brought up in adjoining rooms in the same, big old house. They grew up in the same neighborhood, went to the same Catholic church, and attended the same parochial schools. Yet Charles always had a strong appetite for thrills, adventure, and the unknown, whereas Michael preferred peace, quiet, and the familiar. In interviews, neither complained about lacking anything, and both were positive about the future. When asked about the differences between them, they laughed and said they always had been different, since they were boys. Asked if they would trade lifestyles, both vehemently said no. Each liked and respected the other brother, but there was no way Charles could live like Michael, or vice versa.

Charles got satisfaction from living fast. Michael was more contemplative. Charles was gregarious and always surrounded by friends. Michael preferred solitude and evenings home with a book. Yet both described themselves as content. Their satisfaction did not come from things outside them—the things they did or the people they saw—but from within. The real difference between them was not how they spent their time, but how their brains reacted to what they did. Charles and Michael were fundamentally different at the deepest levels of their personalities. Clearly, something was pushing Charles to try new things and rewarding him for doing them. He did new things because he *liked* novel situations and new challenges; he found pleasure in them. Michael, on the other hand, was rewarded for not experimenting. He built boundaries for himself and stayed inside them because he *liked* familiar things. Something was happening inside their brains to reward them for certain types of behavior.

The temperamental difference in Michael and Charles has been called a variety of names by psychologists.[1] The term used by Robert Cloninger, a psychiatrist at Washington University Medical School in St. Louis, is novelty seeking, which he breaks down into exploratory excitability, impulsiveness, extravagance, and disorderliness. Marvin Zuckerman, the first psychologist to make an extensive study of the trait, called it sensation seeking, which comprises four related but distinct elements: thrill and adventure seeking, experience seeking, boredom susceptibility, and disinhibition.

Whatever it's called, the trait refers to seeking novel sensations and experiences and the willingness to take risks. High novelty seekers find pleasure in varied, new, and intense experiences. They are not necessarily fond of risk, but they are willing to take risks for the reward of the new sensation. Low novelty seekers prefer familiar, conventional, and less intense experiences. They are not necessarily averse to risk; risk is irrelevant. They take comfort in the familiar, which entails no risk.

Although novelty seeking is a single temperamental trait, it can be expressed in many different ways. Physical thrill seeking includes the desire to participate in dangerous sports such as mountain climbing, surfing, or skydiving. Studies have shown that expeditionary climbers, parachutists, and ski instructors are higher than average for this aspect of novelty seeking, while volley ball players and joggers score low. Zuckerman even showed a direct relationship between a person's novelty-seeking score and the speed he drives.

Experience seeking is a second dimension of the trait. Novel stimuli don't necessarily have to be physical, they can be mental or social. New sensations can be found through the mind and senses, such as through avant-garde music and art, exotic travel, or counter-culture experiences. High scorers get excited about new ideas; they are unconventional or innovative. Low scorers are resistant to new ways of thinking; they are conventional. High scorers are eager to meet new people, but not necessarily to get to know them well. Low scorers prefer old acquaintances, even if they don't like them much.

Novelty seeking also works in less obvious ways. Low scorers prefer simple and symmetrical designs: pyramids, crosses, or other geometrical shapes. High scorers like complex and asymmetric shapes, especially those suggesting movement. When it comes to nature paintings, low scorers like calm, tranquil scenes such as quiet lakes. High scorers want the drama of Constable's *Seascape with Clouds*.

High scorers can't bear repetitive experiences, routine work, or boring people. They will never watch a movie twice. High scorers get bored so easily that they will change the way they do things to avoid monotony, even if the change doesn't improve performance. Low scorers won't change even if the current method doesn't work.

In one peacetime experiment, U.S. Army soldiers were asked to volunteer for a hazardous combat exercise. The men who volunteered were the ones who scored higher for risk taking. These would be ideal for commandos. Low scorers, on the other hand, might be better suited for an army in garrison that is more interested in order and discipline than in killing people and breaking things. For example, low novelty-seeking soldiers had the best performance when asked to carefully monitor the position of objects on a computer screen, skills that would be valuable for sentries or radar operators. They were less deterred by boredom and were good at repetitive tasks.

Disinhibition and impulsiveness are the final dimensions of novelty seeking and are the most important to the real-life problems of drinking, drug use, risky sex, and gambling. High scorers cannot control their impulses. Motto: Live fast, die pretty. Low scorers are better able to delay gratification. Motto: Early to bed, early to rise. High scorers live at the edge, spending their money and their energy to the max, consequences be damned. Low scorers are more frugal with their resources, both monetary and psychic.

High novelty seekers have many strengths. Christopher Columbus, Lawrence of Arabia, and John F. Kennedy were probably high for this trait. High scorers go where no one has gone before, act spontaneously, and live life to its fullest. However, living on the razor's edge means it's easy to fall off, and impulsiveness can be costly. The key character traits needed to balance this temperament are impulse control and planning. High scorers literally need to look before they leap. Since the willingness to take risks for pleasure comes directly from the fast-acting limbic system, high scorers need to take a deep breath before acting to allow the cerebral cortex, the planning part of the brain, to play its role.

Low novelty seekers have their own strengths. Queen Victoria and Dwight Eisenhower likely were low novelty seekers. Low scorers are prudent, reflective, frugal, and orderly—and perfectly content to remain that way. They won't be interested in developing a risk-taking, thrill-seeking personality style, because new stimuli give more discomfort than pleasure. They won't be Rambo because they prefer to be Clark Kent. If low scorers feel their temperament is too cautious, that fear or restraint prevents them from enjoying life or having relationships, the problem may not be too little novelty seeking but too much of a separate trait known as harm avoidance. This trait causes fear, anxiety, and worry.

Novelty seeking influences work style and relationships. High scorers have short attention spans and make quick decisions, often without complete information. Low scorers are more reflective and analytical; they can stay focused and want complete information before making a decision. High scorers are quick tempered and show their anger when they don't get what they want when they want it. Low scorers are more even tempered and slower to anger.

A person who scores extremely high for thrill seeking could work as a pilot, firefighter, stock broker, or bank robber. Even a moderately high scorer needs excitement and will be bored by repetitive or routine tasks. They are best in jobs with frequent new challenges and projects, rather than daily routine. High novelty seekers are better talkers and persuaders than listeners and order takers, which means they are better off starting their own businesses than joining large corporations.

Low scorers tend to be orderly and precise, and a routine can be comforting rather than confining. They are cubicle dwellers. Accountants, librarians, editors, machine tool operators, dentists, and computer programmers are good career options. They prefer jobs with long-term projects and goals rather than rapidly changing priorities. They will feel more comfortable at IBM than at a startup company that might not exist next week. Lower thrill seekers make excellent middle managers because they are willing to perform the difficult, often thankless tasks required to turn a new idea into reality.

The confusing thing about trying to measure yourself or others for novelty seeking is that it's possible to score high on one facet and low on another. What matters is the average of all categories. For example, there are people who love jumping out of airplanes but don't get their kicks from sex, drugs, or rock and roll. Likewise, there are people with boring jobs who enjoy avant-garde art. But averaging a large number of people, the fighter pilots like to party, and the people closing down the college library every night are more likely to become midlevel managers than secret agents.

THE PLEASURE CHEMICAL:
DOPAMINE

High and low novelty seekers don't differ in their desire to feel good—everybody likes to feel good—but in what makes them feel good. High scorers need excitement for the brain to feel good. The same level of arousal makes a low scorer feel anxious. A steady, predictable situation would bore a high scorer but comfort a low scorer.

Much of the early thinking about novelty seeking was based on the principle of optimal arousal. Supposedly there was some level of stimulation that caused maximum brain pleasure; too little or too much was thought to be uncomfortable. The theory was that a person with a high score needed a "hot" stimulus to reach the magic level, while low scorers already were close to their ideal level and didn't need so much outside stimulation. Another version of this theory was that high scorers were slow to arousal and needed a bigger jolt; low scorers were more sensitive and needed less stimulation.

Scientists have pushed beyond this rather fuzzy idea of "brain arousal" and have identified the chemicals responsible for pleasure and rewards. One of those molecules is dopamine, the chemical released during good sex, after a delicious meal, or with cocaine and amphetamines. Dopamine is an activator of behavior. It energizes people to seek out things that feel good, and it causes pleasure when those things are found.

Deep in the brain there is a region called the nucleus accumbens that is rich in neurons that produce and respond to dopamine. The nucleus accumbens is the brain's "G-spot," a pleasure center, and the release of dopamine in this area feels good, very good. To find out just how good, scientists placed tiny wires into the nucleus accumbens of rats. They trained the rats to press a lever, which sent out a tiny electrical current. The rats pressed the lever and had their brains tickled. The rats liked that. They liked it so much that pretty soon they did nothing but push the lever, even in preference to pushing levers for food and water. The rats became addicted to the pleasure lever, addicted to self-stimulation.

Then the rats were given drugs that blocked the brain's natural dopamine action. Soon the rats stopped pushing the pleasure lever and returned to more typical rat behavior. The thrill was gone. Without dopamine in their brains, the lever was just a piece of metal.

Dopamine belongs to the group of brain chemicals called monoamines, a family of neurotransmitters involved in many different aspects of behavior—personality, depression, drug and alcohol use, aggression, eating, and sex. Chemically, dopamine is very simple. It is tyrosine, a common amino acid found in many food stuffs, with a few little changes at one end. Dopamine is synthesized in the cell bodies of neurons located in the middle of the brain. One group of cells shoots their axons into the limbic system, the primitive region of the brain involved in emotions, and into the prefrontal cortex, which is involved in reasoning. Other cells work in a part of the brain that gives dopamine the power to activate us to search and explore.

Dopamine alone is not enough to give a rush. Dopamine is a key that opens a lock. The lock is called a receptor, a large protein that sits on the surface of brain cells. The receptor is recognized by dopamine but by no other chemical, just like a lock can only be opened by the correct key. When dopamine snuggles into the waiting receptor, the tumblers turn. Inside the brain begins a series of chemical reactions.

THE DOPAMINE CONNECTION

If seeking new sensations feels good to some people, and the release of dopamine is one way the brain feels good, it makes sense that dopamine might be related to novelty seeking.

To find out, scientists looked first at rodents. The rodent version of novelty seeking, called "exploratory behavior," was measured by putting a rat in a strange cage and watching. Some rats checked out every inch; others sat there, grooming themselves, oblivious to the new surroundings. The rat's movements were tracked with video cameras or with beams of light that recorded how many times the rat crossed a particular spot. A high-scoring rat busily explored the cage. To determine whether the busy rat was curious or just afraid, the droppings were counted. A scared rat left more droppings, which were counted by some lucky scientist who calculated the rate of fecal pellets per minute.

The next step of the research, using mice, was to change the dopamine levels in the brain. Instead of simply using drugs, the scientists actually "knocked out" key genes that regulate dopamine. In a few months, new colonies of mice were bred that were genetically engineered with specific levels of dopamine. In the mice with extra dopamine, in which the dopamine remained "hot" 100 times longer than in normal mice, the result was dramatic. The mice tripped the light beam five to six times more frequently than normal mice and were in a perpetual state of hyperactive exploration. They didn't need any "outside" stimulation to move. Even in an empty cage, they behaved as if they had just found themselves in the most interesting place imaginable—or taken a hit of cocaine.

The reverse experiment produced the opposite result. Mice engineered without the enzyme required to make dopamine were so lethargic that they spent most of the time sitting in the middle of the cage doing nothing. Two weeks after they were born, they couldn't be bothered to eat, drink, or groom themselves. They died of starvation.

The researchers wondered whether the dopamine-free mice were so lethargic because they were physically unable to eat. To find out, some of the mice were injected with L-DOPA, the substance the mice were unable to produce for their own dopamine. Within a few days the L-DOPA-injected mice made a miraculous recovery. They started eating, drinking, and moving around at normal levels. This meant that the lack of dopamine wasn't causing abnormal brain development that prevented eating. Rather, the mice were normal, except they didn't feel like eating. This meant their behavior was being regulated by dopamine.

Sadly, nature has given humans a version of this condition. Parkinson's disease is caused by a degeneration of the dopamine-producing cells in the substantia niagra. The resulting decrease in dopamine production leads to a limitation of movement and trembling of the hands. Doctors who treat Parkinson's disease have long observed that the physical symptoms are accompanied by a personality change that makes people appear serious, stoic, or quiet. When Parkinson's patients were given Cloninger's personality questionnaire, they scored low on novelty seeking but normal on the other personality traits. This lack of novelty seeking was specific for Parkinson's and not a general result of physical disability, because patients with severe arthritis or orthopedic problems scored normal.

Further evidence of dopamine's role in Parkinson's came from treating patients with the L-DOPA used on the mice. As described by Oliver Sacks in *Awakenings,* these patients had been "frozen" and unresponsive since they were young adults. When they were given the chemical needed to make dopamine, however, "this resulted in a dramatic 'awakening' . . . Initially they showed euphoria, interest in their environment, heightened sexual interest, and other signs of increased sensation seeking . . . But as the renewed dopamine output began to act on supersensitive receptors, many of the patients developed manic psychoses followed by depressive 'crashes.' "

The next step in the novelty-seeking research was to look at twins, who can show if a trait is caused more by genes or by some other factor. The key measurement is correlation, which tells how closely two things are related. A correlation of 1.0 is a perfect match, while a score of 0.0 means no relationship. Correlation does not refer to individuals; it's a mathematical way of expressing variances in populations. So in this case, a correlation of 0.0 would indicate that, as a group, twins are no more alike than any random pair of people. A score of 1.0 would mean that, as a group, the twins are exactly alike. When 442 pairs of twins raised together were measured for novelty seeking, the scores for identical twins, who share all their genes, were correlated at 0.59, whereas the scores for fraternal twins, who share only half their genes, were correlated at 0.21. Further analysis calculated that 58 percent of the variability in sensation seeking was due to genes. The remaining 42 percent of the variance came from unique environmental influences experienced by one of the twins and not the other, and measurement error.

The Minnesota study of twins raised apart showed similar results. Identical twins had a correlation of 0.54 for novelty seeking, while fraternal twins scored 0.32. This gave a heritability estimate of about 59 percent, indistinguishable from the estimate for twins raised together.

So regardless of whether twins were raised together or apart, genes seem to account for half or more of the person-to-person variation in novelty seeking. Family environment and upbringing appear to have little if any effect.

A THRILL-SEEKING GENE

By 1995, all the information was at hand to unravel the mystery of the O'Rourke brothers. We knew that the obvious differences between Charles and Michael reflected the deeply rooted temperamental trait of novelty seeking; that individual differences in this trait were largely mediated by several different genes; and that at least some of the genes were likely to be involved in controlling the ebb and flow of dopamine in the brain.

All that remained was to discover the actual genes, which did not turn out to be so simple. Of the hundred thousand or so genes that make us human, there are probably several hundred involved in dopamine signaling and thousands more involved in the rich mixture of sensory, emotional, cognitive, and motor activities that contribute to something like novelty seeking. But the immense amount of information coming out of the Human Genome Project, an international research effort to map all human DNA, was about to enable a breakthrough in the understanding of novelty seeking—and of personality in general.

The breakthrough came from an unexpected quarter: a small research team in Israel led by Richard Ebstein, the laboratory director at the S. Herzog Memorial Hospital in Jerusalem, and Robert Belmaker, an Israeli psychiatrist interested in schizophrenia, which is thought to involve severe dopamine disruptions. Looking at one after another of the dopamine-related genes potentially involved in schizophrenia, they noticed one gene that seemed peculiar. It was the gene that makes the D4 dopamine receptor called D4DR.

The extraordinary thing about the D4DR is that it's so different from one person to the next. This is because right in the middle of the D4DR gene there is a hypervariable sequence of DNA consisting of a string of 48 base pairs that code for 16 amino acid building blocks. Different people have different numbers of this sequence, anywhere from 2 to 11 copies. One version of the gene might have two units of

48 base pairs—96 base pairs—and 32 amino acid residues. But another version of the gene might have 9 copies—432 base pairs—and 144 amino acids. The most common form of the gene has 4 copies of the sequence, and the next most common has 7 repeats; but there also are versions with 2, 3, 5, 6, 8, 9, 10, and 11 copies.

The variation that is passed from parents to their children is enough to alter the function of the D4DR. Different forms of the D4DR were found to have different abilities to bind chemical analogs of dopamine. The longer the protein, the weaker the binding. So differences in this gene were likely to influence how people felt.

The Israeli scientists realized the variation in the D4DR gene couldn't possibly be a cause for schizophrenia because the variation is very common and schizophrenia is rare. It seemed more likely to be involved in a common characteristic that varies from one person to another. Another clue to the function of the gene was that it is heavily expressed in the limbic areas of the brain, the ancient region associated with emotional responses to stimuli. This meant the trait would be at the very core of human behavior. A logical place to start looking for the gene's role was novelty seeking.

The Israeli scientists did a simple experiment. They rounded up 124 subjects from the local university and health center—students, health care workers, and friends—just ordinary people, not patients with mental diseases. The volunteers were given Cloninger's questionnaire, a "true-false" quiz used to measure novelty seeking and other traits. The researchers took a little blood, prepared DNA, and measured the length of the key region of the D4DR gene.

The final step was to see if the length of a person's D4DR gene was related to the score for novelty seeking. It did. People with one or two copies of a long version of the gene, containing six or more repeats, scored on average 0.5 standard deviations higher for novelty seeking than did people with only the shorter forms of the gene, a statistically significant result.[2] The longer the gene, the more the person claimed a desire for new and exciting experiences.

The length of the D4DR gene had no effect on the other personality traits of harm avoidance, reward dependence, or persistence. Moreover, the relationship between D4DR and novelty seeking was apparent in both men and women and wasn't affected by the age, race, ethnic group, or education level of the subjects.

This was exciting news, the first hint of a connection between a personality trait and a specific gene coding for a known protein. The evidence, however, was less than completely convincing. For one thing, the Israeli scientists had looked at only one group of rather modest size. Many previously claimed links between genes and behavior could never be repeated by other scientists using different subjects. A more serious concern was the possibility that the subjects just happened to have different variations of the gene and different levels of novelty seeking.

This kind of error can happen easily. Suppose a team decided to look for a "chopstick gene." They go to Tokyo and Indianapolis and ask people whether they eat with chopsticks. The difference in chopstick usage between the Americans and the Japanese is astounding, off the charts. Next they take DNA samples and find a strong, very significant association between one particular genetic marker—say a blood group gene—and the use of chopsticks. Again there is a statistically significant difference between the Tokyo residents and the people from Indianapolis. Eureka! Since the people with this gene are more likely to eat with chopsticks, that must mean the gene is involved with the ability to use chopsticks, perhaps something to do with eye-to-hand coordination. The scientists pop champagne and publish an article heralding the discovery of a gene for the "successful use of selected hand instruments (SUSHI)." The correlations could be very strong and other investigators likely could repeat the experiment and get the same results. And yet the conclusion would be utterly wrong.

The flaw is this: the SUSHI gene actually codes for something that just happens to have different frequencies in Asians and Caucasians, as is often the case for blood proteins. And Japanese people use chopsticks more than Hoosiers for purely cultural reasons. Therefore the apparent association is spurious.

THE BETHESDA CONFIRMATION

The only one way to *prove* that a gene directly affects a complex characteristic like novelty seeking is to look within families. This is because families are homogeneous racial and ethnic units; and by looking at children raised in the same household, the effect of different environments—including parenting, and the social and cultural background—is minimized.

My laboratory at the National Institutes of Health in Bethesda, Maryland, had been gathering just the right data to test the Israeli findings. During several years we had collected DNA samples and personality questionnaires from a broad range of people who had volunteered for studies conducted at the National Cancer Institute and, in collaboration with Jonathan Benjamin and Dennis Murphy, at the National Institute of Mental Health. The best thing about our subjects was that they included families, mostly pairs of brothers, not genetically unrelated people like the subjects in the Israel study.

Would the Israeli link between novelty seeking and D4DR hold up in our very different group? We had the DNA samples and results of a similar questionnaire to the one used in Israel. The rest was mathematics.

When the computer finished comparing the DNA with the scores for novelty seeking, the results were almost identical to the Israeli study. People with long versions of the D4DR gene scored on average 0.4 standard deviations higher for estimated novelty-seeking scores than did people with the shorter versions. Because our study population was considerably larger than the Israeli population—315 subjects versus 124 subjects—the results were statistically even more convincing. Not surprisingly, one of the subjects with a long gene was Janice, the vivacious blond described in the Introduction, whose high level of novelty seeking was reflected in varied sexual partners and a job in the up-and-down real estate market. Her classmate Ralph, with the same wife and same job for years, was one of the subjects with a shorter version of the gene.

So now we knew the result could be replicated, but was it real? Was the D4DR gene directly related to the difference in novelty seeking, or was it just a spurious association like the "chopstick gene"?

The critical experiment was to look at the families. We focused on those in which one sibling had a long gene and the other had a short gene. If the theory was correct, then the brother with the long version should score higher for novelty seeking; but if the theory was wrong, then there'd be no difference inside the families. In other words, if the theory was wrong, stockbroker Charles O'Rourke could score high for novelty seeking and his brother Michael O'Rourke could score low, but their forms of the gene would be the same. That would mean genes weren't responsible for making Charles a hard-charging windsurfer and Michael an aspiring elementary school teacher.

When the comparison was completed, the result was clear. Charles had a long version of the D4DR gene, whereas Michael had a short version. The same was true of the other pairs of brothers and siblings. The difference was statistically significant, even larger than the difference between all long-gene and short-gene individuals. Even though these brothers were dipped out of the same gene pool and grew up in the same environment, they were different in a recognizable way, and their genes for the D4 dopamine receptor were different as well.

The complete agreement of the two studies was all the more impressive in view of their differences. The subjects came from different ethnic groups: Ashkenazic and Arabic Jews versus non-Jewish Caucasians, Hispanics, Asians, and African Americans. They came from different lands and cultures. Even the type of questionnaires used was different. The concordance of results indicated that the D4DR gene was affecting something fundamental about human nature—not just something peculiar about one specific population, or something limited to one particular way of measuring personality.

After these results were published, another research group found that the D4DR gene was not related to novelty seeking in a small group of Finnish subjects. However, because the Finns are an unusual genetic isolate, it's not clear how relevant this finding will be to the population at large. Recently another group, working in Canada, confirmed the association between the D4DR gene and positive emotions, which is one aspect of novelty seeking, and moreover discovered a connection between the gene, the personality trait, and the density of gray matter in one particular region of the brain. Now the race is on to discover how the dopamine receptor gene wires this part of the brain to be rewarded by novel stimuli.

The Limits of Correlation

Although the connection between the D4DR gene and novelty seeking was clear, it was not absolute. We had not found the single on-off switch that makes some people skydivers and others librarians. Despite the fact that the average novelty-seeking score was significantly higher for long genes than short genes, there were long-gene people who scored low and wild and crazy people who had the short form of the gene. Calculations showed that the D4DR gene accounted for about 4 percent of the total variation in the trait, while the rest of the difference must come from other genes, environmental factors, and measurement error.

Since twin studies have shown that novelty seeking is about 40 percent heritable in a group like ours, and D4DR accounts for roughly 4 percent out of that 40 percent, then this one gene's effect on novelty seeking is about 10 percent. Thus D4DR might be just one of ten different genes that influence interest in novel stimuli. The rest of the genes aren't known yet, but many scientists are looking for them.

The bottom line is that the D4DR gene says something about the probability a person will be a high or low novelty seeker, but it alone is not enough to predict the score.

THE ROLE OF ENVIRONMENT

Part of the reason that no single gene can predict the level of novelty seeking, or of any complex trait, is that "environment"—meaning everything that isn't inherited—also plays a role. This can range from purely biological factors, such as the level of hormones a fetus is exposed to in the womb, to unique experiences, such as a childhood injury or a special third-grade teacher. It includes upbringing as well as social and cultural factors.

It's unlikely that either genes or the environment, by itself, could produce the large difference in novelty seeking seen in Charles and Michael. More likely is that experiences and genes reinforced one another as the brothers developed their adult temperaments. For example, maybe both boys liked to bounce on the bed, until Michael fell off and landed on his face, sending him to the hospital emergency room. Or maybe the same Halloween goblin that scared the wits out of Michael had the opposite effect on Charles: he got a delicious rush of excitement from being frightened.

Although it's not clear if the environment makes the trait stronger or weaker, it's easier to see how it might affect how the trait is expressed. Consider a high novelty seeker born into urban poverty. His curiosity might be expressed in "exploring" the neighborhood with a handgun. The same trait in someone born into a wealthy family might lead to explorations of the commodities market. In the movie *Trad-*

ing Places, the street hustler played by Eddie Murphy swaps roles with the trader played by Dan Akroyd. It's not such a stretch to imagine how one personality type can thrive in different environments. Eddie Murphy's character has a "natural" talent for hustling, and when he's thrown into a wealthy environment, he prospers.

THE EVOLUTION OF THRILLS

At the individual level, novelty seeking is neither "bad" nor "good." For example, a high novelty seeker may be more likely to discover hidden treasure or make a killing on the stock market, but he is just as likely to lose it all on a bad bet. A low novelty seeker may never hit the jackpot and may keep his money in low-yield CDs, but neither is he likely to die in a scuba diving accident or lose everything on a risky venture. On an individual level, if a person is happy, the score for novelty seeking doesn't matter that much.

At the level of species, novelty seeking was probably helpful under some circumstances and maladaptive under others. A high novelty seeker may have been more likely to discover new, more fertile lands or to chase game across the savannah, but he was also more likely to die of exposure or in the claws of an angry lion. A low novelty seeker might have stuck around the cave, but he may have noticed that a particularly tasty or nutritious type of plant always reappeared at the same spot at a particular season or that the tough kernels of a wild grain could be converted to food by diligent grinding and sifting.

Probably more important, a high novelty-seeking man

may have ensured that his long D4DR gene was passed on by chasing a variety of sexual partners. A low novelty-seeking woman might have been just as successful at passing on her short D4DR gene by patiently caring for her children. This ying and yang of novelty seeking may well explain why there is still such great variation in the D4DR gene in modern day humans. It may also be part of the reason why different ethnic and racial groups, who evolved under different environmental circumstances, have noticeably different frequencies of the different variations.

The Thrill of Love

Novelty seeking, like all personality traits, also influences relationships with other people—who you choose to be with, and how you get along.

In some popular songs about love, "opposites attract," but in terms of novelty seeking, the rule is "birds of a feather flock together." Luckily this is so, because studies have shown that having different levels of novelty seeking is a frequent source of dissatisfaction in relationships and marriage.

Behavioral geneticists call "birds of a feather" by the fancier name "assortative mating." There is good evidence of assortative mating because of social attitudes, such as religious beliefs, tolerance for diversity, and political views, and to a lesser extent for intelligence. However, for most personality traits there is no assortative mating: correlations of personality scores between spouses are usually close to zero. Novelty seeking is an exception to this rule. Studies in the

United States, the Netherlands, and Germany have shown that husbands and wives tend to have similar levels of novelty seeking.

To see whether novelty seeking had an impact on happiness in love, researchers used Zuckerman's test to study college students who had been dating for three months or more. They found a strong relationship between similar novelty-seeking scores and contentment with the relationship and sexual satisfaction. It didn't matter if the novelty-seeking scores were high or low as long as they were the same. The lovers who matched for novelty seeking were relatively more likely to be happy and sexually satisfied. The study further suggested that this effect got stronger with time. At the beginning of the relationship, the high-scoring partner might not be dissatisfied with a low-scoring partner because there were so many new things to learn. And the low-scoring partner might have been attracted to a high scorer's "exciting personality." But as time passed, the high scorer became frustrated with the low scorer's lack of enthusiasm for new things, including new sexual behavior, and the low scorer was bothered by the high scorer's unpredictability, including all those "weird" positions and a roving eye.

Despite the evidence, people with opposite scores still get married. Often they end up in counseling. In one study, the novelty-seeking levels of couples in marital therapy were much more divergent than the scores of couples who were more satisfied with their marriages. The further apart the scores the greater the dissatisfaction. Most often the lower-scoring partner is the one who complains and seeks therapy. The low novelty-seeking person is frustrated by the inconsistent, often inconsiderate behavior of the high scorer, so he or she is the one who always wants to "work on the relationship." This is logical since low scorers don't like change, including divorce. Perhaps the high-scoring partner doesn't seek therapy because he or she is getting satisfaction (sexual or otherwise) outside the relationship or is not so upset about junking the whole thing and moving on to greener pastures. In support of this theory, divorced people have higher average novelty-seeking scores than do married people.

The worst combination is a high-score woman with a low-score man; such couples report a high frequency of sexual problems, such as loss of desire and impotence. A high-score man with a low-score woman is less problematic, perhaps because the woman can rationalize that her high-scoring male partner is "just acting like a man," whereas high-scoring women go against conventional notions of "how women should act."

If you should fall in love with an extremely high scorer, get rid of the illusion that you can change him or imagine that he will mellow with age. If he wants to bungee jump, pack his lunch and double his life insurance. If she wants to go dancing with the girls while you stay home, tell her she looks beautiful and walk her to the car. She's more likely to return if you let her go. Don't ever delude yourself into thinking, "Now that he has met me, he'll lose interest in others." Remember that what attracted you to this person is the same thing that drives you crazy. Get used to it, or get out.

On the other hand, if you should fall in love with a low scorer, value this person for the stability, security, and loyalty they bring to a relationship. You might fantasize, "If only she would just try jumping out of an airplane, I know she'd love it." Forget about it. She'll probably get airsick even before you reach altitude.

Novelty seeking not only influences long-term relationships but also brief, purely sexual encounters. High scorers (appropriately enough) are interested in having a variety of sexual partners, like to try out new sexual activities, and view sex as a pleasurable game. Low scorers have fewer partners, practice a more limited and traditional repertoire of sexual behaviors, and see sex as an expression of commitment. New research shows that the D4DR dopamine receptor gene plays an important role in the different sexual behavior of high and low novelty seekers.

ARE YOU A THRILL SEEKER?

By now you are probably trying to figure out your own level of novelty seeking. Here's a simple way to find out: if you are closely reading every word of this chapter, curled up by the fire and sipping hot cocoa, you probably score low. If you are browsing in a book store, with half an eye out for potential sexual partners, and you skimmed all the science stuff to get to the part that best describes your personality, you probably score high.

More likely you seem to be both high and low. Perhaps you prefer a daily routine, but you like jazz. Maybe you are adventurous about sex but not about money. The reason is that novelty seeking, like all major temperamental traits, is a continuously distributed aspect of normal human personality. Everyone has some degree of novelty seeking—the only question is how much. Most people will be pretty close to average, which is why it's called average. In general, though, people who score high are curious, impulsive, extravagant, enthusiastic, and disorderly. Low scorers tend to be indifferent, reflective, frugal, orderly, and regimented.

The difference is all in your mind; what you make of it is up to you.

TWO

WORRY

Seeing the World Darkly

*We are, perhaps, uniquely among the Earth's
creatures, the worrying animal.*
—LEWIS THOMAS

Sally had been painfully shy for as long as she could remember. Growing up, she clung to her mother's skirts and didn't speak unless spoken to, and then only with difficulty. She had few friends as a girl and no boyfriends as a teenager. She was so uncomfortable at school that every morning she threw up her breakfast. When she got older, her first job was at a large bank, where she quickly reached a plateau that was beneath her abilities but felt comfortable, and she remained at the same position for 18 years. She never married, and the only big change in her adult life came when she began caring for her aged parents, an obligation she resented deeply.

At age 41, Sally described herself this way: "I feel angry and hurt most of the time. I feel like my spirit has been shattered and fragmented with each piece having been trampled on and bruised. I am very, very anxious. I am afraid of everything, even centipedes and roaches. I keep thinking something very, very bad is going to happen to me, some great misfortune . . ."

Sally finally sought help from Dr. Peter Kramer, who describes her in his book *Listening to Prozac*. Sally is a vivid example of someone with a painfully high level of a temperamental trait called neuroticism, emotional sensitivity, or harm avoidance. The term harm avoidance is somewhat misleading because the trait doesn't just mean avoiding harm; it refers to a prickly anxiety and a deeply negative worldview that colors everything, a fear of life itself. For someone with a very high level of harm avoidance, life is dark, the future is grim, and every day is a drag.

The reason that some people feel this way is not necessarily because life has too many challenges, or because they were mistreated or abused as children, or because they are weak or lazy. Instead, harm avoidance is a lifelong emotional disposition that is deeply rooted in the genes. Having a high level of harm avoidance is like being born with sunglasses that darken the view of yourself and everything around you. Researchers have found the first signs of harm avoidance even before a child is born. A fast fetal heart rate is a sign that an infant will probably be fidgety and whiny, which often means a high level of harm avoidance in adulthood. By contrast, a low fetal heart rate is a sign of an infant more likely to be smiley and cooey, a way of being that often matures into a rosier view of life.

Harm avoidance is one of the most fundamental, diverse, and persistent dimensions of human psychology, and it has long been suspected of having deep biological roots. Some 1,800 years ago, the Greek physician Galen called it melancholy and blamed it on yellow bile. Today the psychiatrist Robert Cloninger blames it on the brain chemical serotonin.

Sally's score for harm avoidance would be off the chart. A person who scores high is apprehensive, fearful, prone to worry, nervous, tense, and jittery. He or she anticipates harm and failure, especially in uncomfortable, new, or difficult situations. But the anxiety doesn't have to be tied to a specific event or situation; it is free-floating worry. Since high scorers always feel nervous and uptight, they are reluctant to try anything different, to meet new people, or vary from their comfortable routine. By contrast, a person who scores low is calm and relaxed, has a positive attitude, and typically does not worry about the future. The low scorer is uninhibited, nonchalant, carefree, confident, calm, and secure, even when the going gets rough. He or she enjoys meeting others and adapts well to changes in routine.

Sally is also described as showing many of the classic signs of depression, such as tearfulness and poor concentration. This isn't surprising because different elements of harm avoidance, such as anxiety and depression, often go together. Depression is not the same as anxiety or fear, though, and its outward signs are different. Sally also complained of exhaustion, which could be explained by the stress she faced and the fact that depression often disrupts sleep. Such bone-deep tiredness, known as fatigability, is characterized by low energy and the frequent need for naps or extra rest. The high-scoring person recovers slowly from minor illness and stress, while the low scorer is energetic and dynamic, always on the go, and able to bounce back quickly from illness or mental exhaustion.

After a lifetime of feeling bad about herself, what finally drove Sally to seek professional help was anger and resentment at being forced to care for her aging parents. Her hostility was actually a "normal" response for someone like her; while everyone around her seemed to be getting on with their lives, she felt stuck. The easiest people to blame were those at hand, her parents.

Just as depression is an expression of dissatisfaction with oneself, hostility is an expression of dissatisfaction with others. Depression is anger directed within, hostility is anger directed without. Again, surveys show that angry hostility is often linked to other aspects of harm avoidance, such as anxiety and depression. This facet refers to feeling angry, not necessarily acting angrily, which falls more under the trait of aggression. A person who scores high feels angry, bitter, frustrated, and suspicious of others. A low scorer is easygoing, slow to anger, and trusting of others.

Harm avoidance is a blanket trait that includes anxiety, fear, inhibition, shyness, depression, tiredness, and hostility. To a degree, each aspect of harm avoidance is distinct. For example, it's possible to be nervous without being sad, or to be hostile without being tired all the time. Nevertheless, many studies have shown that people who experience one of these negative moods are more likely to experience one or more of the others.

The common feature of all harm avoidance is emotional sensitivity, a kind of emotional sunburn that is easily inflamed. Every little thing makes a high scorer feel bad, pushing out any positive emotions. They are very sensitive to punishment, always waiting for the other shoe to drop. A high level of harm avoidance often is accompanied by an inability to control cravings and urges for food, cigarettes, or possessions. Desires are perceived as irresistible, even if they later produce remorse.

Harm avoidance takes a toll on physical health, causing recurrent headaches, muscle soreness, and digestive problems. Some doctors dismiss these symptoms as "psychosomatic," but they are very real to the people who suffer them. Harm avoidance can even be fatal. In a study of young cancer patients with the same medical prognosis, pessimists were more likely to die sooner than optimists.

A person's level of harm avoidance will have a definite impact on careers and relationships, especially at the high end. Just as a physically frail person would be wise to steer clear of physically demanding jobs, a person with a high level of harm avoidance most likely will shun occupations that require cutthroat competition or jobs that require unfailing cheeriness and constant contact with the public. Low scorers can take advantage of their self-confidence and interest in people, working in sales, public relations, contracting, hands-on health care, everything from cashier to corporate CEO.

As the score for harm avoidance rises, the likelihood of strong personal relationships declines. Harm avoidance does not appear to be a dominant factor in picking a spouse, so there is a good chance your partner scores differently than you do. If your partner scores higher than you for harm avoidance, you might be unfairly blaming yourself for his or her unhappiness. High scorers need patience, affection, praise, and support. Just because your partner doesn't share your bubbly enthusiasm for life doesn't mean he or she wants to change, however. Nor does it mean you should suppress your own good feelings. In fact, your partner probably is attracted to how easily you express joy.

GOING TO THE SOURCE

Where does harm avoidance come from? A bad life or a bad outlook on life?

In Sally's case, probably both. Kramer describes a traumatic childhood, financial setbacks suffered by her parents, and a cramped home where young Sally witnessed her parents having sex. Some analysts would say that was enough to force a sensitive child into a pattern of behavior like Sally's. But what made her a sensitive child in the first place? Why did she respond so severely to her surroundings, while other people raised under much tougher circumstances grow up to be outgoing and confident?

Research on hundreds of children has shown that harm avoidance is one of the earliest expressed and most persistent aspects of personality. Louis Schmidt, a graduate student in Nathan Fox's lab at the University of Maryland, is one of the researchers tracking harm avoidance in young children. Asked about his research at the Institute for Child Study, Schmidt gladly played a videotape.

The tape opens with three little girls walking into a friendly playroom decorated with colorful pictures. Drawing paper, crayons, and toys are on the carpeted floor. One girl, whom we'll call Rhonda, takes the lead.

"Hey, crayons!" she says and runs to the pile of stuff on the floor, plops down, and begins to draw.

She is joined by two other girls, also about four years

old, and all three begin to chat easily. A bit later, the door opens to reveal a girl we'll call Valerie. She practically has to be dragged into the room by her mother and never lets her back leave the wall. Reluctantly, she makes a half-hearted attempt to open the box of crayons, but when her mother leaves, Valerie bursts into tears.

The more confident Rhonda asks, "Are you scared? We're not scared."

The painfully timid Valerie doesn't stop crying, even when Rhonda offers a fistful of crayons. "We have to do something for this girl because she's crying," Rhonda says. Rhonda even tries to help by complimenting Valerie's shoes, but the shy girl heads for the door and futilely turns the knob to escape.

An adult appears and leads the girls in a story-telling game. The bold Rhonda has to be told to sit down when her turn is finished; she wants to keep talking. The two other girls play politely and tell stories about their birthday parties, stopping when their turns end. The shy Valerie can't be coaxed to open her mouth. She is sitting cross-legged, hunched over, eyes firmly on the ground. A white blanket is clutched tightly in her hands, and she continuously rubs it back and forth on her face. When the woman asks her to stand up, she mumbles something indistinct. "Come on, honey," says the lady. "Don't you want to tell us just a little about your birthday party?"

Valerie buries her head in her lap and lets out a great, heaving, heart-wrenching sob, and dissolves into tears.

"That's okay," the woman says, comforting the girl while signaling to her fellow scientists operating the hidden cameras that this experiment has ended for now.

The four girls in the playroom are part of a large study

trying to identify the earliest signs of temperament and eventually to understand how this inborn temperament interacts with upbringing and environment to unfold into personality. These girls were chosen because they show the full range of temperament called "inhibition," what most of us call shyness and appears to be the childhood version of harm avoidance. The shy Valerie obviously is extremely inhibited, and the bold Rhonda scores at the opposite end of the scale. The other two girls are "normal," scoring somewhere between Rhonda and Valerie.

The fascinating thing about these four-year-olds is that when they were only four months old, they scored exactly the same. Videos of Valerie as an infant show the same timid and fearful behavior. Left strapped in a car seat, Valerie reacts strongly to a series of things designed to stimulate her. When a recording plays nonsense syllables such as "bo bo," she at first seems raptly intent. Then she makes a scared, fearful face and cries. When colored mobiles are placed nearby, she whimpers. When her mother finally appears at the end of her first "experiment," Valerie frets and fusses until picked up.

Rhonda, on the other hand, was as bold at four months as at four years. On the videotape, the nonsense sounds make her laugh, the mobiles have her cooing. At the sight of her mother, Rhonda lights up with a beaming smile. She squirms as much as Valerie in the car seat, but it seems to be more in pleasure than in discomfort.

Five months later, when the girls were nine months old, they were connected to an EEG machine to measure brain waves. Even before the machine was switched on, Rhonda and Valerie reacted differently to being fitted with the little caps wired with electrodes: Valerie fussed and Rhonda seemed to enjoy the attention. Not only that, their brain waves were different. Valerie showed more activity on the front right side than on the left, while Rhonda showed a more active left side.

This difference wasn't just a fluke; Fox and his University of Maryland colleagues have found a consistent pattern of more right frontal activation in inhibited or shy children and more left frontal activation in uninhibited or bold children. The pattern holds true both when the children are resting calmly and when they are under stress. The correlation continues into adulthood: highly anxious adults show more right frontal activity than less anxious individuals. The right frontal adults are also more likely to show signs of depression.

This pattern of brain waves fits with what is known about the role of the right and left frontal areas in the control of emotional expression. The right side seems to be more involved in controlling negative emotions, whereas the left side plays more of a role in positive emotions. The anxiety felt by Valerie may be because the more dominant side of her brain, which is the right side, is sending out more negative signals than the left side, which sends out the positive signals.

Other physical tests show similar variations. For example, the saliva of the shier girls has two times more of the stress hormone cortisol than that of the bold girls. Other measurable differences include how quickly they blink in response to a loud noise and how their pupils dilate under stress. The shy kids have "tighter" muscles, especially in the face, and have a harder time when asked to make faces. The shy children show a greater increase in blood pressure when they stand, and although the resting heart rate is similar for all, the shy children show bigger increases in heart rate under stress, both from physical activity and tough mental tasks.

Over the next year and three-quarters, the girls made two more visits to the lab. On one tape, Valerie scrunches down into her mother's lap and barely looks up, even when a stranger tries to engage her in play. As soon as a clown comes into the room, Valerie loses it completely and begins to cry. When the same scenario is played out for Rhonda, she also scoots to her mother's lap when the stranger enters, but she soon is on her feet playing with a new friend. At the appearance of the clown, Rhonda becomes tense and quiet, but within a few minutes she recovers her composure and plays easily with the clown.

Fox and Schmidt were not surprised to see that Valerie's shyness seemed as deeply imbedded in her personality as Rhonda's boldness. The reason is that inhibition, what we call harm avoidance in adults, is one of the most enduring aspects of temperament. This is particularly true for children like Rhonda and Valerie, who were chosen because they are on extreme ends of the spectrum. Most children are somewhere in the middle, although even among average people lifetime patterns of shyness are clearly visible.

Lifetime shyness or boldness has been confirmed many times by Harvard University's Jerome Kagan, the granddaddy of such research. Kagan, who was the mentor of Fox, studied children who were classified as very inhibited or very uninhibited when they were two years old. When they were seen again at seven and a half, more than three-quarters of those initially classified as inhibited scored above average for inhibition. The shy seven-year-olds stayed at the edge of a group of other children, they read or played alone instead of joining group play, and they were quiet and stared at other children without speaking. Three-quarters of the initially uninhibited children scored below average for shyness five years later. The bold seven-year-olds were more likely to initiate activities and to be talkative, noisy, and boisterous.

The same behavior patterns, now more sophisticated, were evident when this group was seen again at 12 to 14 years of age. During interviews with a child psychiatrist, the inhibited children spoke only when spoken to, but the uninhibited ones often came back with their own questions. The shy children were grim and dour as they slogged through the questioning, while the bold ones were quick with smiles and laughed frequently.

Kagan, Fox, and their colleagues believe that shyness has important roots in the amygdala, an almond-shaped mass in the limbic system, the brain's emotional response center. The amygdala tells the body what to do when something bad happens. It takes in information from the senses—such as a loud noise or a burned finger—and triggers a racing heart, rising blood pressure, a startled jump, and a fearful face.

Kagan, whose own initial belief in the importance of the environment was tempered by his findings of inborn shyness, thinks that inherited differences in the amygdala account for the differences between shy and bold children. Much of the evidence for this theory comes from animal experiments. When the amygdala is destroyed, laboratory animals become more tame and lose their fear of threatening things. By contrast, when the amygdala is artificially stimulated with electricity, the animals show signs of fear and agitation. They are also more likely to develop illnesses such as gastric ulcers.

In humans the best evidence for the biological roots of shyness is that it can be measured in so many physical responses, from fetal heart rate to hormones and brain waves. The problem is that the observations can't say whether the stressed heart beats faster because the children are shy, or whether the heart rate has something to do with causing the shyness. Maybe it's just a coincidence. One of Kagan's explanations is dubious. He found that shy children were more likely to have light-blue eyes than bold children, which he asserts is a direct relationship between eye color and temperament. He speculates that it has to do with the production of the brain transmitter norepinephrine and melanin, the chemical that makes some eyes brown. This is a theory spun out of thin air. The supposed relationship between shyness and eye color could be just a spurious association, as irrelevant as saying that shyness is related to time of birth or to the obstetrician's first name. The more logical explanation is that northern Europeans who frequently have light-blue eyes also happen to have higher levels of anxiety for some completely unrelated reason. At any rate, Fox and Schmidt have not replicated the association between eye color and shyness in the children they have studied.

This is not to say that shyness isn't biological, but a better way to search for biology's role is to look directly at genes.

THE ROOTS OF WORRY

One of the hallmarks of genetically influenced traits is that they run in families. In his study of the desperately unhappy Sally, Kramer found that her parents were themselves quiet and prone to depression and physical ailments. They described Sally's emotionality as a family trait, common to her mother, grandmother, father, and sisters. Kramer was not surprised; there are many famous examples of harm avoidant traits running in families. An extreme case would appear to be the Hemingway family, where depression has ended in suicide too many times to be a coincidence. Ernest Hemingway, who committed suicide at age 61, is thought to have used the same shotgun that his father used to end his own life 33 years earlier. Ernest's brother and sister also committed suicide. Margaux Hemingway, one of Ernest's granddaughters, struggled with bulimia and alcoholism. She was found dead during the summer of 1996, at age 41.

The histories of the Hemingways and many other families suggest that genes play a role in depression and other aspects of harm avoidance. But just because something runs in a family doesn't necessarily make it genetic. If that were true, religious preference, favorite recipes, and surnames also would be "genetic." Once again, a more systematic way to investigate the role of genes is by studying twins. Such studies show that genetic influences on harm avoidance appear early and remain in evidence for a long time. For example, one study of twins by Adam Matheng at the University of Louisville found that identical two-year-olds were more similar than fraternal twins in their reaction to a stranger. Whenever identical twins are more similar than fraternal twins, the likely reason is genes, because identical twins are genetic clones. Another study, this one at the University of Virginia of 350 pairs of seven-year-old twins, estimated that 50 percent of shy, fearful, and inhibited behavior was inherited.

The results are strongest for the extremes of shyness and boldness. Scientists at the Institute for Behavior Genetics in Boulder, Colorado, found that 50 to 60 percent of shyness was inherited in twin babies. But when they focused on the most bold and most shy children, the heritability rate rose to

70 to 90 percent—one of the highest recorded rates for any aspect of behavior and probably the reason it doesn't change much during a lifetime.

Even in adulthood, after all the life events that would seem to make a person more inhibited or less so, harm avoidance is still largely influenced by genes. A large number of studies have shown that genes account for about 40 percent of the variation in several measures of adult harm avoidance. The rest of the variation is almost all due to "unique environment," the experiences that are not shared by family members. "Shared" environment, which includes general parenting style, social and economic status, the school system, neighborhood, etc., has relatively little effect on harm avoidance. People are born inhibited or something unique happens to them that doesn't happen to siblings. Whether parents push all their kids—or pamper them—doesn't seem to matter as much in terms of making them more or less inhibited.

Twin studies also are being used to determine if the different aspects of harm avoidance—such as depression and anxiety—tend to be seen together because of the same genes or the same experiences. In one of the first of a new kind of analysis, called a multivariate study, Ken Kendler and his colleagues at the Medical College of Virginia looked at anxiety and depression in 3,798 twin pairs, both identical and fraternal. Similarly to other researchers, they found that the separate symptoms of anxiety and depression were 33 to 46 percent heritable. But when they looked at the coinheritance of the two traits, they got a surprise. Genetically the two traits were more than 99 percent the same. It seemed too simple to be true, so the experiment was repeated, looking at the clinical diagnoses of generalized anxiety disorder and major depression. Again, the number for the genetic covariance between anxiety and depression came out at nearly 100 percent.

Finding close to 100 percent inherited covariance between personality factors or psychiatric symptoms is truly astounding. It means the exact same genes that cause anxiety are also responsible for depression. This doesn't mean that a given individual has to be both anxious and depressed. Rather it means that one set of genes can make a person either anxious or depressed or, less frequently, both. The end result of the same set of genes—depression or anxiety—depends on nongenetic factors, such as life events. For example, the death of a close relative often causes depression, but usually not anxiety. The illness of a relative increases anxiety more than it does depression. Thus anxiety comes from the anticipation of loss whereas depression comes from the experience of loss. What brain mechanism could make a person extrasensitive to loss?

PROZAC POINTS THE WAY

When Sally came to see psychiatrist Peter Kramer, he prescribed what was then a new drug—Prozac. Within two months, the most obvious sign of Sally's depression, her weepiness and inability to cope, had disappeared. After four months, Kramer reported, ". . . she looked brighter, calmer, self-assured, in firm control of herself. The most important effect of the medication, Sally felt, was that it cleared her head—made her more awake and aware, more confident of her perceptions." In Sally's words, "The medicine helps me to clarify problems. It takes me less time to find positive solutions. I don't panic. I don't feel my 'brain hurts' under stress, and I don't obsess."

Prozac worked for Sally like a silver bullet because it targets the brain chemical serotonin. Like dopamine, serotonin is a monoamine, a small simple molecule. It is synthesized from tryptophan, an amino acid found in foodstuffs such as milk. The reason people drink warm milk to sleep is because milk is rich in tryptophan, which is converted to serotonin, which can aid relaxation. Milk may even ward off nightmares because substances that excite serotonergic neurons suppress dreaming. By contrast substances that inhibit serotonergic transmission, such as LSD, increase dreaming. LSD has a structure very similar to serotonin and can bind to serotonin receptors but does not activate them; LSD is like a decoy that distracts the receptors and keeps them from functioning.

The serotonergic system is the most widespread neurotransmitter system in the brain. The cell bodies of the serotonin nerve cells are clumped together in the raphe nuclei deep in the midbrain, and the axons branch out like veins throughout the brain. They spread extensively into the limbic system, the "heart" of emotional responses. They wind into the cerebral cortex, which is involved in cognition and sensory perception, and into the frontal lobes, which are involved in impulse control, empathy, and social awareness. Other branches reach the hippocampus, the site of memory and learning, and the hypothalamus and pituitary gland, which are involved in appetite and sex.

The extensive branching of the serotonin system makes it a blunt instrument. Instead of conveying highly specific information, the system is better at firing up large sections of the brain. Because of serotonin's broad-brush effect on the brain, anything that alters the system will have a big impact on mood, self perception, and behavior. Factors that can influence the system include drugs, life experiences, and, from the beginning, genes.

Genes give the body the blueprints for the serotonin system and then build it. The system is in place before birth, responding to outside stimulus and the body's needs. When one of these neurons is excited, packets of serotonin inside the cell migrate to the wall of the cell and spill their loads into the spaces between other cells. Two things can then happen to the released serotonin: it can bind to receiver molecules called receptors, which starts a chemical reaction, or it can return unused to the packet, called reuptake. The reuptake is controlled by a special type of protein called a transporter. The transporter is the target of Prozac.

If levels of the serotonin transporter are low, or if it doesn't work well, then more serotonin is left splashing around the brain, jolting everything it touches. If the transporter works well, most of the serotonin gets sucked back up before it jangles the brain.

There are many different serotonin receptors, each of which is produced by its own distinct gene. Already more than a dozen distinct serotonin receptors have been identified and their genes have been cloned. This diversity of receptors, along with the widespread geographic distribution of the serotonin nerve cells, is why serotonin affects so many different brain functions. But there is only one serotonin transporter, which comes from just one gene. So anything that affects the serotonin transporter will affect all of the psychological traits controlled by serotonin.

For years people suspected that serotonin was involved

in harm avoidance. The evidence included the fact that low levels of serotonin metabolites were found in the spinal fluid of people who had attempted suicide or took their own lives. Also, lowered amounts of serotonin transporter protein were found in severely depressed individuals. The problem with this evidence, though, was that scientists couldn't measure the actual free serotonin levels or the amount of serotonin signaling in the brain, so it might have been an indirect correlation.

Another clue pointing to the role of serotonin was the success of the first popular antidepressant drug, Iproniazid. The drug, originally developed to fight TB, inhibits the cleanup of serotonin and forces the cells to release more chemicals, increasing the number of cells firing. A side effect is that it makes people feel good. The evidence that serotonin was involved in feeling good or bad wasn't solid, though, because Iproniazid cleaned up everything, not just serotonin. Maybe it was acting on another brain chemical that affected mood.

The next major antidepressant to be developed was Imipramine, which affects serotonin by a different mechanism: it inhibits reuptake, or the effort of the cells to recapture spilled serotonin. This drug also seemed to be proof that serotonin was linked to depression, but Imipramine also affected other brain chemicals, which accounted for its many side effects.

The search was on to find a drug that would narrowly target serotonin and nothing else. Scientists at Eli Lilly laboriously synthesized one molecule after another to find one that would block the uptake of serotonin into cells. Compound 82816 was the payoff. In animal experiments, the drug blocked serotonin uptake 200 times better than anything else and without so many side effects. Known as fluoxetine, the new drug was sold under the brand Prozac.

Prozac was invented to help with major depression, which is defined by the American Psychiatric Association's *Diagnostic and Statistical Manual of Mental Disorders* as a series of symptoms such as persistent depressed mood, loss of pleasure or interest in life, psychological or physical agitation, sleep disturbances, weight change, loss of energy, and difficulty thinking and concentrating. This type of depression is called "major" because it becomes the most important fact of the patient's life.

The good news was that Prozac was effective in treating major depression. The bad news was that it wasn't that much better than Imipramine. The real breakthrough came when Prozac was applied to other conditions, such as minor or "atypical" depression, generalized anxiety disorder, panic disorder, and social phobia. In all these cases, people reported relief when using Prozac. The interesting thing about this list of problems is that it neatly parallels the most extreme forms of harm avoidance. If Prozac worked on serotonin *and* helped people deal with harm avoidance, it seemed likely serotonin was responsible for bringing together the diverse aspects of harm avoidance.

The problem with this theory is that it's not yet clear how Prozac works. The Eli Lilly explanation is that Prozac inhibits the serotonin transporter, which in turn increases the amount of free serotonin in the brain. This theory assumes that low serotonin is what causes depression and raises harm avoidance. Therefore, high serotonin should decrease anxiety and depression. But another school of thought, supported by experiments on animals and humans, holds that serotonin signaling is what causes anxiety and depression. According to this theory, Prozac must ultimately decrease serotonin signaling. This may be why the effects of Prozac typically require weeks or even months of treatment, even though the drug inhibits serotonin reuptake within a matter of hours. While the evidence so far shows that serotonin plays a key role in harm avoidance, it's not clear whether the difference between being happy or sad is caused by low serotonin, high serotonin, or changes in the level.

The effects of Prozac on the less severe aspects of harm avoidance can be amazing. People not only are able to get over major depression, they feel great. For example, once Sally's depression was gone, her personality and life continued to change on Prozac. Kramer wrote that Sally not only improved her situation at work by becoming more assertive, but she also started dating and going to dances. Sally talked glowingly about how Prozac let her true personality shine through for the first time in her life. After 18 months on the drug, she became engaged.

"This is a big step," Sally said about her coming marriage, "but I feel good about it. I am moving along fast, but not too fast. I love him, and he loves me. Before, I only felt closed in; now I feel happy."

GENETIC PROZAC

Sally was able to change from sad to happy simply by taking a drug that altered her serotonin. But we still don't know why she was sad in the first place or why other people seem naturally happy. If genes are the key, the first place to look is at the gene that produces the serotonin transporter, the Prozac target.

When the transporter gene was first isolated, it sparked a frenzy of experiments in which people tried to prove that it was the golden key able to unlock the mysteries of any number of mental illnesses. The initial results were discouraging. There was no link to manic depression, nor to schizophrenia, nor to any other of the standard psychiatric diagnoses. The golden key didn't fit any of the locks. Soon the psychiatric geneticists gave up on the serotonin transporter and moved on to the next gene-of-the-month. This is an example of why psychiatric genetics is regarded as a faddish science: experiments are as much influenced by what's "hot" as by what's logical. In this case the shrinks-turned-scientists threw up their hands and said, "Nothing there." Even if the transporter was somehow important to behavior, they decided that all people must have pretty much the same version.

The psychiatric geneticists were wrong, and it wasn't the first time. Early on, they had erred in the opposite direction: they kept "proving" things that weren't true. First they announced they had found genes for schizophrenia, manic depression, alcoholism, and other conditions. Later they couldn't replicate the results or they were proved incorrect. These "false positives" received a lot of press, but more insidious are the many "false negatives," meaning claims that a particular gene is *not* involved in a disease when perhaps it is. In a way, false negatives are worse than false positives because they cause people to stop working when they could be on the verge of a breakthrough.

Fortunately, there were two scientists working on the serotonin transporter gene who were not so fickle: Dennis Murphy, a senior NIMH researcher who has been studying serotonin for more than 20 years, and Peter Lesch, a former postdoctoral fellow with Murphy who now heads his own laboratory at the University of Würzburg in Germany. They were particularly interested in how the gene was turned on and off by signals from the body such as hormones and stress. They went beyond the gene itself to look at bits of DNA upstream that promote the activation of the transporter gene. There was one region of the gene that was especially intriguing because it had an unusual composition and structure. Also, when this bit of DNA was removed, the activation, or expression, of the gene increased. That meant the region was designed to slow down the process of turning on the gene. Intriguingly, the region contained 16 imperfect repetitions of the same sequence of 21 to 22 bases. It was like a piece of music repeated 16 times with slight variations. Lesch suspected that this repeated structure might be important because such sequences were often different from one person to the next.

When Lesch looked at this bit of DNA in different people, he indeed found a difference. About 57 percent of the genes had the complete 16 copies of the repeated sequence, but 43 percent of them had a shorter version with only 14 copies, a difference of 44 bases of DNA. It was as if the DNA in some people had gotten bored of singing the same tune and stopped a little early.

There were three ways to see if this variation made a difference. First Lesch and his team shot the DNA fragments into cells growing on a petri plate. The long version was expressed twice as strongly as the short version. To be sure the DNA wasn't acting differently out of its natural environment, they tested white blood cells with either the long, the short, or both varieties of the gene. Cells with two copies of the long gene made about twofold more serotonin transporter RNA than did cells with one or two copies of the short version. The final experiment was to see how much serotonin transporter protein was being made. Sure enough, cells with only the long version of the gene captured significantly more serotonin than cells with a short version.

The three experiments showed that everybody has a gene that makes serotonin transporter, but they make different amounts. About 32 percent, or one-third, of the population have two copies of the longer, more powerful form of the gene and therefore make high levels of serotonin transporter. The remaining 68 percent, or two-thirds, have one or two copies of the shorter version of the gene, which is dominant, so they make less of the transporter.[1]

THE GENE AND PERSONALITY

Murphy and Lesch's persistence had paid off. They'd found an inherited variation in DNA that clearly affected serotonin transport. Here was a little button effectively controlling levels of serotonin in the brain. Now the key question was, What did this "genetic Prozac" do in human beings? What effect did it have on temperament?

Murphy and Lesch were pretty sure what the gene was not. They doubted it was an on-off switch for some mental illness; it was just too common. If it were a controlling factor in major depression, for example, that would mean at least one-third of the entire population should be feeling terrible. They figured the variation played a role in a more normal—meaning more common—variation in personality. The problem was that most of their research subjects were uncommon: they were psychiatric patients with serious problems. The researchers needed DNA samples and personality profiles on a broad range of healthy people. They called me.

Murphy and I had collaborated previously looking for the novelty-seeking gene. He knew I had exactly the material he needed, and within a few weeks, postdoctoral fellows Sue Sabol in my lab and Dietmar Bengel in Murphy's lab had genotyped 505 individuals for the DNA region just upstream from the serotonin transporter. They looked at every person for whom we had personality scores—college students from local campuses; gay men from our studies of sexuality and AIDS; brothers and sisters, fathers and mothers; young and old; white, black, yellow, and brown; male and female. Once all the bench work was done on the DNA, it was a simple matter to match the data with the personality test scores.

We held our breath as the first results rolled down the computer screen. Given all the various functions ascribed to serotonin—from anxiety to depression to aggression, not to mention eating and drinking, cognition, and sex—it was an open question whether we'd see anything specific or indeed anything at all. Maybe serotonin was responsible for so much that it wouldn't be linked with any of the narrow traits we were looking for. On the computer screen, we were searching for stars. As the statistical results were churned out, two stars (**) indicated that a result was significant at the $p < 0.01$ level, which meant less than a 1 percent chance of being a fluke.

The earliest results were based on a standard personality test that measures five major traits. First we ran the DNA data against the factor for conscientiousness, which means dependability and organization. If the people who scored high or low for conscientiousness shared anything in common about this bit of DNA, the computer would catch it and mark it with a star. The first results were negative; no stars, no correlation. Next we ran the data for the trait of openness; nothing. I crossed my fingers as we punched in the third factor, extroversion. The numbers ran, the results popped onto the screen. No stars.

The first hit came with agreeableness. That made sense because one aspect of agreeableness (or the lack of it) is aggression, a trait ascribed to serotonin. The correlation was weak, though, only one star. It wouldn't be much of a link, but it was encouraging. The computer ground on. The fifth and final factor was neuroticism, a measure of anxiety, emotional stability, and reactivity to stress. These were exactly the traits that should be involved if we really were looking at a genetic Prozac. If we didn't get a hit here, we weren't going to find it anywhere.

I watched the screen. New numbers popped into view. My face lit up in a big smile. Not one star. Not two stars. But three stars—less than 1 chance out of 500 that the correlation was just by chance. And right bang on where we expected it.

This was great news, but it was only a beginning.

Next we started looking at the data from every possible angle. Maybe there was something obvious we were missing, or perhaps the way we had collected the subjects had biased the sample. First we split the subjects into those who'd been recruited through the National Institute of Mental Health and the National Cancer Institute. There was no difference: both groups showed the same significant correlation to neuroticism but not to the other factors. Next we checked females versus males, and straights versus gays; the neuroticism factor continued to shine through. We corrected the data for age, ethnic group, education and income; no matter how we crunched the numbers, the result stayed the same.

I still wasn't convinced. It was possible the results were just some sort of coincidence based on the five-factor structure of the personality test. The questions we asked people about their personalities or how we categorized the answers could have led us in the wrong direction.

Fortunately our subjects had been given not one but two different personality tests. Since there is much disagreement about how to describe and measure personality, we wanted to use as many different yardsticks as possible. The second test is called the 16 Personality Factor inventory, developed by the pioneering American psychologist Raymond Catell in the 1940s. The 16PF divides personality into five factors formed from 16 core traits.

When we matched the DNA data on our subjects with Catell's traits, a star appeared by only one of the five superfactors: anxiety. The correlation was right where we expected it to be, confirming that we really had found a link between the DNA region and a basic personality trait.

The final analysis was to look at Cloninger's predictions. He had theorized that serotonin would be involved in harm avoidance, and now we had a chance to test the theory. He was right. We found a significant correlation between this DNA region and the trait he called harm avoidance, which we could estimate by mathematically rearranging the questions from the five-factor test. There was no correlation for any of the other traits he identified. This was the second confirmation of Cloninger's theory. He also had guessed right that dopamine was linked to novelty seeking, and now he was proved right that serotonin was linked to harm avoidance. His model of personality was starting to look pretty good indeed.

There was one intriguing twist to the story. The people with the highest level of anxiety-related traits had the short version of the gene promoter. This meant that where the serotonin transporter was least efficient, people had the most anxiety. This was the opposite of the Eli Lilly explanation for how serotonin works; it should have been the lowered serotonin transporter levels that were associated with decreased harm avoidance. Our results were more consistent with the "classical" model that serotonin causes rather than alleviates anxiety, depression, and other elements of harm avoidance.

It's difficult to draw any firm conclusion yet about the direction of serotonin action, however, because it's possible that a lifelong decrease in serotonin transporter gene expression actually decreases serotonin signaling through a feedback or compensatory mechanism. What's really needed is a direct way to measure serotonin signaling in the living brain—but that's not available yet.

Our study not only confirmed the connection between the serotonin transporter and harm avoidance, but it also provided the first conclusive evidence that the multiple facets of harm avoidance are connected at the level of the genes. That's because the differences in the DNA correlated equally well with several different aspects of harm avoidance: anxiety, depression, hostility, pessimism, and fatigability. Thus the results were a satisfying confirmation of the claim that a single set of genes—in this case just one gene—can influence distinct traits that are obvious in real people.

THE EVOLUTION OF ANXIETY

Why should two-thirds of the population have a gene variation that causes them to feel anxious, depressed, and pessimistic? Is Mother Nature really that cruel? Obviously the most extreme forms of harm avoidance don't make evolutionary sense. If the goal of genes is to be passed on to future generations, then a severe phobia that causes people to be so afraid of social interaction that they are unlikely to mate is counterproductive. The same is true for depression so severe that people are too lethargic to have sex or care for children. But the gene for the serotonin promoter region is more often associated with mild anxiety and depression than with severe forms, which strange as it may sound, is a good thing.

Evolutionary psychologists argue that mild anxiety and depression are useful under many circumstances. For example, a fretful, crying baby is more likely than a calm, quiet baby to attract a mother's attention and care. One study in Africa showed that infants with a "difficult" disposition fared better than those with an "easy" temperament during famine. Likewise anxiety can be a useful reminder of life's perils. An early hunter who turned and fled at the lion's roar was more likely to hunt another day than one who felt no fear and attacked the lion with bare hands; just as today a person who avoids bad neighborhoods at night is less likely to be killed than someone who goes looking for trouble. A sense of disappointment and sadness when goals are not met might allow a person to stop, think things through, and devise a better plan—instead of blithely continuing with a course of action that doesn't work.

Such arguments may well explain why the harm-avoidant form of the serotonin transporter gene still exists, but it doesn't really explain why it exists at such a high level. Our research, which is described in more detail in Chapter Six, suggests a more direct mechanism: sex. It turns out that people with the short serotonin transporter gene have sex more frequently than do people with the long form of the gene. This will come as no surprise to psychiatrists and sexologists. Mild anxiety is often associated with increased sexual drive, and one of the most frequent side effects of drugs like Prozac is to lower sexual drive.

Depressed or not, happy or sad, the gene doesn't care about how you feel. It's sole concern is to be passed on to the next generation. The only way to be passed on is through sexual relations, so guess what? A gene that makes you anxious and sexually active is more likely to survive than a gene that doesn't.

TILTING THE SCALES

People can be anxious or calm, depressed or happy, angry or serene for many different reasons. The serotonin transporter is one of the reasons, but it's not the only one. Our data showed that differences in the serotonin transporter account for 3 to 4 percent of the overall variability, or about 7 to 9 percent of the genetic variance in the combination of neuroticism, harm avoidance, and anxiety. Thus assuming that the other genes have a similar power, the serotonin transporter might be just one of 11 to 14 different genes that contribute to these aspects of temperament. The other genes aren't known yet; probably some of them are involved in other aspects of serotonin signaling while others may be involved in still unknown pathways.

And of course genes aren't everything. The behavioral genetics studies show that environment—meaning everything that isn't inherited—is at least equally important in harm avoidance. Exactly what the important environmental factors are, however, isn't clear. Kagan has evidence that inhibited infants whose parents encourage them to fend for themselves are more likely to lose their shyness as they grow up than are infants who are mollycoddled. But upbringing can't have very much of an overall effect because twin studies show that shared environment, which includes general parenting style, has almost no statistically discernible effect on harm avoidance.

Scientists aren't even sure which unique life experiences are important. One unique experience—and one that most people would consider positive—is falling into a lot of money. But even a windfall of money isn't enough to tip the scales of anxiety, research has found. Edward Diener, a psychologist at the University of Illinois, found that lottery winners are no happier a year after hitting the jackpot than they were before. The same is true with many of the factors that we usually associate with contentment; getting married, having a family, or a job promotion have only minor effects on happiness.

Equally surprising is that negative events, such as divorce, losing a spouse, getting fired, or even suffering a spinal cord injury, have only minor correlations to depression. Typically such events, either the good ones or the bad ones, make people happier or sadder for a few weeks or months, but within half a year their mood has returned to the previous level. This has led some researchers to propose that the brain has a set point for happiness just as the body has a set point for weight.

The combination of influences explains why there are exceptions to the genetic rule. For example, Russell, 41, has the long version of the gene and should score low for harm avoidance, yet he scores in the top 10 percent. He has had significant problems with depression and is seeing a therapist. Despite an IQ well above average, Russell works at temporary clerical jobs because he lacks self-confidence. On the other hand there is Daniel, a middle-aged man who has the short version of the DNA and should score high for harm avoidance. Yet he scores in the lower 10 percent, he's never seen a psychiatrist, he's happily married, and he leads an active social life. Despite what should be a genetic strike against him, Daniel is a confident, successful, easygoing, and affable architect.

Both men are proof that DNA is not necessarily destiny and that character traits can override genetic predisposition. In the case of Daniel, his strong character worked to overcome what could have been a genetic problem. Even though Russell didn't have a disadvantage at birth, he still was miserable. Just as a person can be tall and not play basketball, you can have "happy" genes and feel terrible. If you feel your own level of harm avoidance is too high, there are plenty of ways to improve your outlook on life, such as working on a positive attitude, changing jobs, losing a few pounds, taking a vacation, exercising, or even indulging in an occasional hot fudge sundae.

If your level of harm avoidance is paralyzing, however, you may want to seek professional help. Some of the most severe manifestations of harm avoidance are seen in bipolar disorder, commonly known as manic depression. Sufferers swing between immobilizing depression and periods of intense activity and euphoria, known as mania. There is an effective medication, lithium, that helps smooth out the highs and lows, although the drug does not work for some people, and others say they miss the "up" periods.

Bipolar disorder is more than just severe harm avoidance. The brain mechanisms are not yet understood, but what is clear is that bipolar disease has a strong genetic component. It runs strongly in families, and systematic studies have shown that the parents and siblings of manic depressives have a 10-fold increase in disease rate. Not surprisingly, it also runs in twins: the identical twin of a manic depressive has nearly a 40-fold increased chance of having the disease.

Obviously environment is also important for bipolar disorder, although we aren't sure what about the environment matters. The best evidence that environment is important is that rates of bipolar disorder have increased substantially during the previous 20 years and that the age of onset is getting younger. Something appears to be changing in the way we live that is increasing the rate of manic depression.

A major global search is underway to find the genes for this crippling condition. Already several large chromosomal regions have been identified, but the actual genes remain hidden. Part of the reason for the mystery of bipolar disorder is that it doesn't seem to make sense in terms of evolution. Why would such a difficult condition survive? Kay Jamison, a psychologist who studies manic depressive illness and who herself has the disease, says that in some people, one positive outcome of the condition is tremendous creativity. Jamison says that many famous artists, writers, and scientists could have had the disease, including Vincent van Gogh, William Blake, Walt Whitman, and Edgar Allan Poe.

Manic depression is an extreme case, but a touch of the blues doesn't have to be a bad thing. All emotions or feelings—even the bad ones—have value and importance. The goal is not to suppress emotions but to balance the negative feelings with positive feelings. Happiness depends on the ratio of positive to negative feelings, not on suppressing all feelings of anxiety, sadness, and fear. Everyone has negative feelings sometimes; without them, life would be isolated and cold. It is only when negative emotions are allowed to get out of control that they interfere with life and can lead to illness.

Although the harm-avoidance score reflects a genetically set mood range, there is considerable flexibility within that range. In other words, it's easy to move above and below your natural level, and much depends on conscious effort. David Lykken, a psychologist who has studied the genetics of harm avoidance in twins, advises, "Be an experiential epicure. A steady diet of simple pleasures will keep you above your set point. Find the small things that you know give you a little high—a good meal, working in the garden, time with friends—and sprinkle your life with them. In the long run, that will leave you happier than some grand achievement that gives you a big lift for a while."

THREE

ANGER

Aggression, Crime, and Violence

Don't push me 'cause I'm close to the edge.
—GRANDMASTER FLASH AND THE FURIOUS FIVE

It's late Friday afternoon. Just as you are clearing your desk to pick up your son at the day-care center the boss calls you into the office. "Where's the Jones report?" he asks. The question is irritating because just that morning you had explained that the report would be delayed because the person you assigned to write it has called in sick—or so she says. The thought now occurs to you that maybe she really left early for a weekend at the beach. Your body stiffens and your face becomes flushed and warm. You are about to snap at the boss, but you bite your tongue. You swallow your anger and say, "It will be done by Monday morning, I promise." As you discuss the report, you realize your weekend is wrecked.

Even though you made an effort to be calm, your body reacted to the stress strongly, even violently. Deep inside your midbrain, the amygdala delivered a dose of quick-acting chemicals that tightened muscles, pumped blood to the surface of the skin, and made you irritable. The same organ, no bigger than an almond, also released messengers to the adrenal gland, causing it to produce stress hormones that made the brain even more sensitive to the next insult or stress. Your body is on high alert, primed for action. Walking out to the car, you try to put the office behind you, but it's too late: your brain is in overdrive.

By the time you get to the day-care center, it's 6:20 P.M., 20 minutes after closing. At $1 a minute, the penalty eats up what you had saved all week by bringing your lunch to work. All the other kids have been picked up, and there's your son standing alone on the curb, clutching a lunchbox and today's artwork. The day-care person appears from around the corner. Normally you are happy to see her and to hear about your son's adventures, but you can't believe she left him out on the sidewalk alone.

"Where were you for goodness sake?" you ask peevishly.

"Where were *you*," asks the young woman, more tense than normal. "We close at six, and I just ran in to call your office to see if you'd left."

You fight mightily the urge to scream at her. Instead, you take a breath and say, "Sorry. It was one of those days. Thanks for waiting."

As you strap your child into the car seat, you realize you have overreacted, but it seems that things are building up: your boss is a jerk, your coworker is lazy, you should not have been late . . . You just want to get home and have a drink. Your son is talking and talking, but you don't really hear above the roar inside your head.

First you have to stop at the supermarket, and of course the parking lot is packed. Finally, as you start to pull into a space, you are cut off by a young man in a beat-up Honda. If looks could kill, he is dead. He is oblivious, until you honk, and then he gives you the finger. You imagine all sorts of clever insults about his parentage, his car, his face, but you keep them to yourself. The aisles are crammed and the chilled air is filled with noise. Your fingers are gripped tightly around the handle of the cart. Inside your head, chemicals are surging. If you were a prize fighter, you would be more than ready to knock the head off your opponent. Instead, you have to concentrate on placing only nine items in the cart so you can go through the express checkout lane and get home.

"Mommy, can I have some candy?" your son asks, using the whiny tone that he knows will get a reaction.

"No."

"Why not?"

"Because."

"EEEEIIII!" he shrieks, squirming and pulling himself out of the cart so he almost falls onto the hard floor.

"Sit down and shush up!" you hiss, pushing him back into the cart. But you don't see his leg is caught, and the harder you push, the louder he screams. You are a fraction of a second away from slapping his face.

"Stop your whining! You are a very, very, bad boy!" He continues to squirm, so you grab his chin roughly, bring his face close to yours, and bark just one word: "Stop!"

People are staring now, and your son knows you aren't playing around. He's rarely if ever seen you like this, but even at his young age, he can read the danger signals. Mercifully, he behaves like a perfect gentleman. You pay for the groceries, pile back into the car, and head for home. As you begin to calm down, you wonder: why did I do that? Why do I let all these little things build up? I almost hit my own child. What was I *thinking?*

The problem was that by the end of this chain of events, you weren't thinking. Instead, the amygdala, part of the primitive limbic system of the brain, had taken over. The cortex, the more modern, thinking part of the brain, was pushed aside and ignored. You were responding the same way a cave-dwelling ancestor would have responded to an attacking mountain lion. The human brain is hardwired to be capable of anger, and in a quirk of our nature, the brain responds to feelings of hostility, frustration, and rage not with reason or judgment but with even more anger. Genetically and biologically speaking, we are all close to the edge— and getting closer every minute.

The question is not whether you get angry sometimes— if you didn't you wouldn't be human—but what you do with those feelings. Do you let your anger progress into aggression and violence? Or do you harness it to work harder, be braver, and perform better? Do you let your anger build and build until you explode, lashing out at whoever is unlucky enough to be in your path? Or do you channel the energy effectively?

Just because anger is "natural" doesn't mean it's pretty—or that you have to give in to it. It is possible to develop emotional and cognitive habits that control anger and harness it productively. All of us try to control our tempers, which is why the regulation of anger has been such an important part of human society. Over the centuries people have devised many ways—individual and collective—to control natural anger. If every individual could control his anger 100 percent, we wouldn't need so many laws—or prisons. We have organized sports to release and reward aggression within proscribed rules and boundaries. Religion also works as a check on natural aggression. Christianity tells us to turn the other cheek, and other faiths teach how to let anger pass through the body. Some people turn to drugs or alcohol to dull the edge of anger. Modern self-help programs focus on managing aggression.

In the case of anger, the advice of the ancients is better than that of the pop psychologists. Loving your enemy, hard as it may be, is actually healthier for everyone. Turning the other cheek *is* better than fighting, and not just in a moral sense. The worst thing you can do, which is actually what some therapists recommend, is to "vent" your anger. This theory is a myth, and practicing it will only escalate the situation and fill you with rage. Another myth about aggression is that angry people are solely the product of a bad childhood or bad neighborhoods, or that loving parents can keep any child from growing up to be violent. These theories are based on a misunderstanding of how the brain works and the nature of human temperament.

Anger is usually directed at someone else, which is what makes it a social problem. The progression from hostility and frustration to antisocial behavior and crime is one of the most important issues we face as a society. Nobody wants to live in a world in which assault, murder, rape, and mayhem are commonplace, yet we are allowing it to happen. Societies break down and wars are fought. People slaughter each other because they come from the wrong tribe, the wrong religion, the wrong country, or the wrong neighborhood. This is not simply the result of "natural" human aggression.

It's obvious that the current U.S. crime wave can't be blamed on biology. The global gene pool today is virtually identical to the gene pool 100 years ago, or 1,000 years ago.

We are the same people genetically; what's changed is the way we live and govern our behavior. Crime is not an individual problem, but a social problem. The way to predict which individuals are going to be violent today cannot be done by testing their blood in a lab. The best way to predict who is going to be violent is to see where and how they live. Today's criminal violence is not about brain chemicals but about poverty, the gulf between rich and poor, racial polarization, urban squalor, a lack of personal responsibility, family breakdown, and the deterioration of civil society.

However, social and environmental factors alone aren't enough to explain violence and crime. If that simple-minded view were true, then everybody born in the ghetto would be a criminal, and everybody born in plenty would be a model citizen. The truth is more complex. All the research shows that anger and hostility—and their visible outcomes such as crime and violence—are caused neither solely by the environment nor by biology. Genes don't make criminals—neither does "gangsta" rap. The mix is what's deadly: the combination of genes and environment, temperament and character. There are individuals who are violent because of what's going on in their heads, and there are societies in which violence flourishes. In no other domain of human behavior are nature and nurture so thoroughly intertwined.

BEYOND NORMAL ANGER

Although everyone gets mad sometimes, there are some people for whom anger is a permanent state. One of them is Megan, a middle-aged woman who runs a small printing and copying service.

Megan doesn't like most people. Either they are stupid or too smart for their own good. Either they are poor because they are lazy or rich because they are lucky. When Megan doesn't like a person she usually let's them know. During the past six years she has had 11 different part-time employees, not one of whom lasted longer than nine months. The most recent quit after being called a "moronic imbecile" for accidently running off a few hundred extra copies of a large printing order. His final paycheck was reduced by the amount of overage.

Even minor annoyances can be very frustrating for Megan. When somebody cuts her off in traffic, she doesn't hesitate to express her anger by leaning on the horn or shouting out the window. More than once she has been threatened back with fists and even a tire iron. Running out of staples is ample reason for a tirade against her employees. As a child, she once threw an entire set of Lincoln Logs down the laundry shoot when she couldn't figure out how to build the house shown on the box.

Megan assumes that other people think of her as poorly as she thinks of them. Indeed her favorite pastime is trying to figure out why people dislike and avoid her. Usually she assumes it's because she's of Middle Eastern descent and a woman in business. The idea that her own behavior might turn off others has never occurred to her. When a customer switches to another business service, or an acquaintance doesn't invite her to a party, she plots elaborate revenge. Forgive and forget is an alien concept. Her motto is remember and retaliate.

There is one thing Megan really enjoys, and that's rowing, a sport she discovered in college. When she was a freshman, competitive rowing was only for males. Megan changed all that when she hired a lawyer and threatened a lawsuit if the college didn't include women. When they gave in, she formed a female rowing team and became its captain. She trained hard; the physical exercise made her feel better. She never did very well in eights or fours because the other women couldn't keep up, and she didn't enjoy taking commands from a cox. But at single sculling she became very adept and won many races. There was nothing she enjoyed more than pulling ahead and seeing all the other boats fall behind her.

Megan is an extreme case, but she illustrates the basic building blocks of anger and aggression. The most obvious facet of her personality is hostility, a pervasive antagonistic view of the world and people around her. In terms of personality systems, hostility is related to harm avoidance, neuroticism, and anxiety. The connection between hostility and these temperaments is not surprising: harm avoidance, neuroticism, and anxiety are about feeling bad; hostility means feeling bad about other people. This is not the type of harm avoidance that makes people shy, but rather the type that makes them see everything—including other people—in a negative light.

Feeling bad about people is one thing, expressing it is another. Psychologists have different ways of classifying anger. It can be a lack of compliance, which means a lack of agreeableness. Robert Cloninger sees aggression as a combination of harm avoidance, a temperamental trait, and a lack of cooperativeness, a character trait. No matter what it's called, aggression is the acting out of hostile feelings—whether through harsh words or physical violence.

Megan has frequently been advised to "count to ten" when she gets angry, but she rarely gets beyond one. Psychologists call this quick trigger-finger impulsiveness, lack of self-control or lack of restraint. What it means is acting without deliberation, and it's a common characteristic of angry people. Coupled with Megan's impulsiveness is an unusually low level of fear. Although we usually think of fear in a negative way, it is one of our most useful emotions. Fear is what keeps most people from doing ultimately self-destructive things such as punching out the boss. Fear is felt in the gut: it is a physical sensation, a tightening of the stomach, drying of the mouth, clamminess of the skin. These are the bodily responses to the signals of the limbic system.

All of us have some degree of hostility, belligerence, and impulsiveness. The question is how much and how well we control it. In Megan's case, her aggressiveness has had one positive outcome, her success at rowing. Angry people typically prefer to compete rather than to cooperate. To them the world is a jungle. Like novelty seekers they love action and adventure, but they are motivated more by the excitement of winning than by risk. They especially like dog-eat-dog competition with only one winner and many losers.

Although Megan's anger and aggression have hurt her and her family, she is in one sense lucky. She has never murdered anyone, smashed up a bar, or been put in jail. She may be obnoxious but she's not a felon. Has Megan kept out of trouble because she was raised by an affluent suburban family, or because she's a woman, or because her inherited temperament does not include violence? What makes the difference between someone who is an aggressive criminal or just an angry nuisance? Is it genes, environment, or both?

DOUBLE TROUBLE

Disentangling the role of genes and environment in a trait as complex as criminality is not simple. Because children are both the genetic and environmental products of their parents, it's difficult to distinguish the cause of behavior. The same parents provide the genes and the environment, making it hard to see which is more important and how. The most useful method is to study children who are separated from their biological parents shortly after birth and are brought up by genetically unrelated parents. If the child turns out to be more like his biological parents, genes are clearly involved. If he grows up more like the people who raised him, environment is key. This kind of experiment happens all the time; it's called adoption.

The most comprehensive adoption experiments in the United States have been conducted by Remi Cadoret and his colleagues, who have studied more than 1,000 Iowa families during the past twenty years. They compared the biological children of parents in trouble—with the law, alcohol, or getting along with others—to the biological children of parents without such problems. All the children were separated from their biological parents at birth or within a few days and were adopted by families with no blood ties. The basic question is: are the adopted children more like their biological parents—who contributed the genes—or more like the parents who reared them—who contributed the "environment"?

The answer is neither and both. For the children with nonproblem genes, the home environment made little difference. Even if the new parents got divorced, abused alcohol or drugs, or had other problems, the children turned out the same as kids raised in good homes. Some turned out good and others got in trouble, but the ratio was normal; it didn't change because of their home life. In other words, when the genes were good, the rearing environment didn't matter that much.

When the genes were bad, however, the home environment meant the difference between success and failure. When the home environment was bad, the children who inherited problem genes were at risk. In these households, the level of childhood and adolescent aggression was dramatically increased. Measures of bad behavior such as lying, stealing, truancy, and school expulsions were up by as much as 500 percent. Some of these children turned out fine, but the ratio of bad kids to good kids was skewed. Aggressive and antisocial behavior increased dramatically only in children with both "bad genes" and bad homes. That shows that what is being inherited is not bad behavior or aggression, but rather a genetic sensitivity to the environment. The genes didn't make them antisocial; the genes made them vulnerable. A bad seed planted in good soil had a decent chance. A bad seed planted in bad soil withered. Humans by nature respond well to nurture, but they also respond to its absence.

A child with temperamentally aggressive characteristics runs a double risk because his or her behavior can actually disrupt the family environment. In one extreme instance, a four-year-old boy from an exceptionally troubled background was adopted by a couple who already had a little girl of their own. Even at this age the boy was a real problem, and a week after bringing him home, the parents called the adoption agency to return him. With counseling they reluctantly agreed to keep him, but his behavior caused many arguments and fighting between the parents, and in two years they were divorced. Eventually they did get the boy out of their home. He shot and killed their daughter and was jailed.

Some of the best adoption data come from Sweden, a country with meticulous records of adoptions, court proceedings, and hospitalization not as easily available to researchers in the United States. The Swedish studies looked at every one of the 862 men born out of wedlock in Stockholm between 1930 and 1949 and adopted at an early age by nonrelatives. The records were searched for information about criminal behavior, drinking, and medical problems. Biological parents, their children, and the adoptive parents were compared to see who was similar, who was different.

The criminal records showed a striking interaction between genes and environment. When both the biological and adoptive parents were from the low-crime groups, the adopted men had a low (3 percent) arrest record, mostly for petty crimes, which was about the same as in the general population. Being adopted into a high-risk family raised criminality slightly, to 7 percent. The next highest were the biological children of the high-risk parents, who had an arrest rate of 12 percent. But by far the highest rate of petty crime—a whopping 40 percent—was seen in those children who had the double misfortune of high-crime biological parents and high-crime adoptive parents. Again, these were bad seeds in bad soil.

Although these results might seem scary to prospective adoptive parents, it's important to note that most of the children ended up fine, regardless of genetic background. If anything, the studies are good news for people thinking about adopting because they show that even the biological children of the worst offenders don't necessarily follow in their footsteps. And for the kids with the worst possible genes, the best possible home can make the difference.

When Genes Matter Most

Good parents can successfully raise a child with bad genes, but timing is everything. Just as a slightly wilted flower can be revived with water, so a child can be helped with love. If you wait too long, however, both the child and the flower are beyond saving. When is the critical period for intervening in the path from bad genes to bad behavior? The answer comes from studying twins.

Twins can be used to determine when the genes or the environment is most important. The reason is that genes are not switches that are only flipped on or off at conception. For example, you may have been born with a gene for baldness, but your hair won't fall out until middle age. Criminality works much the same way: certain genes don't kick in until later in life. In childhood the most important factor in determining who breaks the law is the environment. Later, genes matter most.

The evidence for the changing balance between genes and environment comes from looking at the childhood and adult behavior of the same people, thousands of twins who served in the Vietnam War. In interviews with more than 8,000 of the men, about 9 percent of them showed juvenile antisocial behavior, and about 9 percent showed symptoms of the same behavior as adults, levels similar to the rest of the population.

The importance of genes was measured against two

things: shared environment, which means common upbringing and schools, and unique environment, which is everything else, including measurement error. The following percentages gauge the factors responsible for antisocial behavior as young people and as men. For juveniles, genes were weighted 7 percent responsible, and common environment 31 percent; the remaining 62 percent was due to unique or unknown factors. The results were the opposite for adults. For antisocial behavior among grown men, genes were weighted 43 percent responsible, whereas common environment was at 5 percent. Unique factors accounted for the remaining 52 percent.

What could be responsible for this remarkable flip-flop that makes upbringing more important than genes for boys but less important for adults? Multivariate analysis showed that the common environmental factors, meaning upbringing, were the same in both age groups, but their effect was getting weaker with time. The genes were the same, too, but they were acting more strongly later. The only thing that changed qualitatively over time was the unique environment—those things experienced by one twin but not the other.

Other studies of children and adults have shown a similar pattern. For example, in twin studies of juvenile delinquency conducted in North America, England, and Japan, the correlations were 91 percent for identical twins and 73 percent for fraternal twins. That means that if one twin is a delinquent, his twin brother, whether identical or fraternal, has a frighteningly high chance of also being a delinquent. These rates are so high, and so similar, that they appeared to be measuring a contagious childhood disease rather than delinquency. Bad behavior was spreading through certain households and neighborhoods like chicken pox. Bad environments, it seemed, rather than bad genes, were turning out bad kids. In many of the neighborhoods studied, more than half of the kids got in trouble with the law. In such a rough neighborhood, it would take extraordinary genes to keep a child *out* of trouble.

A very different picture is seen in adults, in whom antisocial and criminal behavior shows a more typically genetic pattern. Seven different studies of criminal behavior conducted since the 1930s in North America, Germany, Denmark, Norway, and Japan found average concordance rates of 52 percent for identical twins and 23 percent for fraternal twins. The fact that the correlations are so different means genes are playing a big role. Looking at the data from the largest study, which was conducted in Denmark, heritability was separately calculated as 76 percent for repeated property crimes and 50 percent for violent offenses against persons. This looks much more like an inherited phenomenon.

The picture that emerges is this: when children are living together in the same home, they are powerfully influenced by their shared environment. If their dad is selling drugs, their friends are in gangs, and their schools are violent, it's hard to keep out of trouble. Give the same child a successful, respected father and good school, and you're more likely to end up with a model young citizen. Once the children leave home, however, they are less influenced—or restrained—by what they learned growing up. They can pick their friends and neighborhood, in essence choosing their environment. It's not their genes that are changing, but the genes help to create their own environment, whether good or bad.

What puts kids in juvenile hall is mostly their environment, but what puts men in jail is more their genes. This means there is an opportunity to intervene in the pathway between genes and criminality, and that opportunity occurs early in life. The trick is to understand what makes the difference, whether it is family or friends or neighborhood, and then to fix the problem.

SEROTONIN'S LASH

U.S. Marines pride themselves on being tough and aggressive, so it's appropriate that the first link between brain chemistry and violence came from studying leathernecks.

The link was found by Frederick Goodwin, a biological psychiatrist who later got into trouble for comparing the behavior of inner-city youth with caged monkeys. (His research is better than his choice of metaphors.) Goodwin's interest in the biochemistry of violence began in 1979 when he published a paper about men being considered for psychiatric discharge from the U.S. Marine Corps. His plan was to study the behavior of the men, and then to look at their brain chemicals. He hoped the "bad" marines would share something common in their brains.

First, Goodwin studied their records for signs of "excessive violence and psychopathic deviance." Then he wanted to know if they shared a unique difference in the chemistry of their central nervous systems. Goodwin and colleagues tapped the spines of the problem marines to analyze their spinal fluid, which is rich in brain messenger chemicals.

Soon they detected a pattern. The discharged marines had decreased levels of one chemical, 5-hydroxyindoleacetic acid (5-HIAAA), a breakdown product of the brain transmitter serotonin. All the other chemicals seemed normal.

Low levels of the same chemical were also found in a broad range of violent or aggressive people, including prisoners convicted for impulsive, aggressive crimes, children who tortured animals, disruptive school children expressing hostility towards their mothers and aggression toward others, and men with unusually high scores for aggression, irritability, hostility, or psychopathic deviance.

The connection between serotonin and violence should come as no surprise. Remember from Chapter 3 that serotonin is thought to be a key player in depression, anxiety, hostility, and other facets of the temperamental trait known as harm avoidance. The relationship makes sense if depression is regarded as a form of anger or bad feeling about oneself, while aggression is viewed as anger toward others. Serotonin's job is to make people feel bad; it's the brain's punishment chemical. Whether people are mad at themselves or at others, whether depressed or aggressive, serotonin is involved.

Once the link between serotonin and violence had been tentatively established in humans, scientists began to manipulate serotonin levels in animals. The results were immediate and profound. Artificially increased serotonin in mice, rats, and monkeys caused them to be less aggressive. Lower levels of serotonin increased aggressive and impulsive behavior.

These results were found not only in pristine labs. J. Dee Higley journeyed to Morgan Island, a 475-acre island off the coast of South Carolina, to study the population of wild rhesus monkeys—a close relative of man. He looked closely to see which monkeys were engaging in aggressive behavior and which ones were avoiding it. Higley and his team also recorded the number of old scars and recent wounds, a measure of how much trouble the monkeys had found.

They divided the adolescent male monkeys into three groups and tapped their spines. Some were particularly aggressive, picked fights, and were covered with scars and wounds; some were especially submissive; still others were in between. The most aggressive monkeys had the least serotonin, the least aggressive had the most. Interestingly, the most aggressive monkeys also had increased levels of two other chemicals, the monoamine norepinephrine and the hormone ACTH, both thought to be markers of stress.

The serotonin study showed that this one brain chemical accounted for upwards of 25 percent of the aggressive fighting behavior in these monkeys. When the scientists graded the monkeys for how well they got along with each other, they found the exact opposite results. The gregarious monkeys, the ones who spent their days socializing and grooming other monkeys, had the highest levels of serotonin.

The next step in serotonin research, using mice, was to find out how the chemical works. The logical place to start was at the serotonin receptors in the brain, where the chemicals transfer information. If the receptors were involved in aggression, then blocking the receptors should change the level of aggression. Genetic engineering allowed scientists to breed a strain of mice totally lacking one such serotonin receptor called 5-HT1B.

The first mice born without the 5-HT1B receptor appeared normal and healthy. All males, the mutant mice lived together for a month before the experiment began in earnest. Another group of normal male mice was raised separately. Then unfamiliar mice were put in the cages. In the cage filled with normal mice, the residents weren't happy with the "intruder." They rattled their tails, and within 160 seconds they attacked for the first time. This was the normal response of natural mice. There were displays of anger and violence, but much of it was just huffing and puffing. In terms of real physical violence, there was fewer than one attack on the intruder every three minutes.

The mouse who was dropped into the colony with the serotonin mutation had a tougher go. He didn't last 90 seconds before he was attacked. He was bit, chased, and threatened six times every three minutes. The attacks were quicker and came six times more often than in the normal colony. To put those numbers in perspective, imagine being dropped into a neighborhood where the violent crime rate is 600 percent above normal. Simply by losing one of the more than a dozen different serotonin receptors, normal mice were converted into crazed killers.

It looked like an open and shut case: low serotonin causes aggression. Since genes have a role in serotonin levels and how it's used, then genes must be causing aggression.

But that didn't explain Mike McGuire's monkeys. Michael McGuire and Michael Raleigh kept vervet monkeys at UCLA. When drugs were used to manipulate serotonin levels, the results were as expected: low levels made aggressive monkeys, high levels made calmer monkeys. The monkeys with the highest levels—those who got along best— were most likely to rise to leadership positions in the group.

The monkeys were regularly visited by school children, and McGuire liked to recruit their scientific expertise. He asked the children to pick out the leaders of the group, and after watching the monkeys a few minutes, the children almost always picked out the dominant male. "If I ask graduate students to do the same thing," McGuire said with a laugh, "it takes six months."

What the children saw was how the monkeys had sorted

out the leadership positions. The dominant male took control, and the rest followed his lead. The experiment began when McGuire and Raleigh reorganized the troops to flip the hierarchy. Suddenly the top-ranked males found themselves on the bottom, and the followers became leaders. When the serotonin levels were checked again, the new leaders had higher serotonin levels than when they were on the bottom. The former leader of the pack became hostile, irritable, and prone to random acts of violence. His serotonin level had fallen. McGuire hadn't artificially changed serotonin levels, just social position. That alone was enough to change the level of serotonin.

A variation of this monkey house experiment also worked in a frat house. Raleigh and McGuire tested college fraternity brothers and found that the leaders had higher serotonin levels than new pledges. Nobody would argue that there is a "frat gene," so the most likely explanation is that a person's place in the pecking order is the cause rather than the effect of different serotonin levels.

If this is happening in society at large, it doesn't take much imagination to realize that a person who is born in poverty, who lives in a slum, who doesn't have a good education, who has the "wrong" skin color, religion, or language might expect to have low social status. With that comes low serotonin; and with low serotonin comes aggressiveness, hostility, and violence, which, of course, can only lead to more of the same. Obviously this tendency to violence occurs only to a small minority, even in the worst environments, so other factors are at work.

MALEVOLENT MALES

The question of whether genes play a role in aggression and crime is tremendously controversial. Yet even the most vocal critics, those convinced that only bad environments produce bad people, can't deny one simple biological fact. The most important single factor in whether a person is violent or aggressive is determined by just one chromosome, the Y, which assigns gender.

The statistical evidence is overwhelming. Men are charged with five times as many aggravated assaults as women, ten times as many murders, and 86 times as many rapes, according to the U.S. Department of Justice. In all history, wars have been waged mostly by men. Aggression today doesn't just get men killed, it also makes them rich and famous. Professional sports are dominated by men, especially men who can channel their strength toward a specific goal. And what do other men like to do? Watch violence on television or in stadiums.

Martin Daly and Margo Wilson, who studied murder around the world for their book *Homicide,* didn't want to be misled by the social and physical dominance of men in most cultures, so they compared rates of men killing men to women killing women. They went to American cities, African villages, towns in India, and communities in Scotland. Every place they looked, the answer was the same. Males commit many more murders than females.

The most impressive finding was that the relative rate of male-to-female homicide remains constant even as the absolute rate varies. In other words, although there is much more killing in an American city than in rural Wales, the same percentage of the murders are committed by men. For example, there were 3.7 killings per million people every year in England and Wales between 1977 and 1986, while during a similar period in Detroit there were 216.3 killings per million people. Yet in both places there were at least twenty times more killings by men than by women. The ratio was 23.1 in England and Wales and 28.7 in Detroit.

If the only difference between men and women is the single Y chromosome, is it directly involved in aggression? Experiments with mice suggested that it does have an important role, since changing the structure of the Y chromosome can alter the level of aggressive behavior. The theory was perfect: the Y chromosome is the difference between men and women, and men are violent, so there must be something on the Y that's involved in violence. The problem was finding a way to test the theory on humans.

Scientists knew that in some very rare cases, men are born with two Y chromosomes instead of one. If the Y chromosome was making men violent, then maybe the men with two Ys would be even more violent. To test this idea, scientists did an experiment that seemed logical, but that in the end presented more problems, both scientific and ethical, than it solved.

First they looked in prisons, and sure enough, the evidence was there. In the general population, about 1 of every 1,000 men has two Y chromosomes, but in certain prison populations the figure was five times greater. In one prison for the criminally insane, the rate of double-Y men—dubbed "supermales"—was 19 times higher than in the normal population. The press went crazy with the story, and some speculated that a "cure" for violent crime was just a test tube away.

The prospects were so intriguing that a group in Boston

set out to track double-Y babies during their lives. If the babies grew up to have a high ratio of aggression or crime, it would be even more proof the Y chromosome was involved. This was technically the right way to design an experiment, but there was no way to do it ethically because baby boys are not mice. The scientists told the parents that their boys had two Y chromosomes, which wasn't fair to the boys and wasn't a good research design because the parents might rear the children differently knowing their condition. When the community learned that little boys and their families were being used as guinea pigs, there were protest marches and demonstrations against "biological determinism," and plenty of academic agony, before the study was shut down. There was one good outcome of the experiment. It is one of the reasons why all universities, medical schools, and research institutes now have review boards to monitor the ethics of proposed research on humans.

After the scandal had passed, there still was no answer to the original question: does an extra Y make men violent? A large and careful study in Denmark provided some help. The study looked at the tallest 15 percent of more than 30,000 men born between 1944 and 1947. The scientists concentrated on the tallest individuals because other studies had shown that an extra Y tends to make men taller. Even in this skewed sample there were just 12 men with two Y chromosomes. This shows that the extra Y syndrome is so rare that it can't possibly be a major source of human aggression. We know there are a lot more than 12 angry men out there, even in Denmark.

But there were some interesting findings about the 12 Danes. First, they were less intelligent than average and more likely to drop out of high school. Second, they had a fourfold increased rate of criminal records. More than 40 percent of them had been arrested, although mostly for petty crimes rather than violent ones. So if there is any connection between the Y chromosome and criminality—and that's still a big question—it seems to be through an indirect effect on intelligence rather than a direct effect on aggression or violence.

THE ANGRY HORMONE

Probably the most important role of the Y chromosome in aggression is manufacturing testosterone, the hormone that makes half the population male.

Part of the proof is that testosterone, which is secreted by the testis, rises sharply at puberty in males—just when levels of aggression rise—and slowly declines during adulthood—just like levels of aggression. But this theory loses some of its brilliance with a harder look. Boys sprout hair on their bodies at puberty—just when they become more aggressive—and men lose hair as they age—just when they lose aggression. If there were any correlation between body hair and aggression, then the world would be more peaceful if men shaved their chests.

A better experiment is to look for a connection between testosterone and aggression in men the same age. In a study of 4,462 U.S. military veterans, the ones who ranked in the top 10 percent for testosterone had significantly increased antisocial behavior: assault, physical aggression, going AWOL, trouble with parents, teachers, and peers. They also had increased drug and alcohol use and multiple sexual partners. Scientists found the same results in hockey rinks and gyms as they did in barracks and prisons. For example, the higher the testosterone, the more aggressive the hockey player. Among 26 judo contenders, the ones with the highest testosterone levels were the most verbally aggressive and the least likely to tolerate frustration. The relationship between maleness, testosterone, and aggression is found in almost every species, not just in humans. This suggests that it is the product of evolution and not something learned. The one exception actually proves the rule.

Spotted hyenas are one of the few species run by females. They live in packs of up to 80 members, and the leader is always female. All adult females are dominant to males. For example, when a pack devours its prey, the biggest strongest male demurs to the smallest female. The females don't maintain their position with sugar and spice; they bite and claw fiercely at the slightest provocation.

Researchers at Berkeley have shown that this unique reversal of sex roles is caused by testosterone. Unlike in most species, all hyena fetuses—male and female—are bathed in the womb with high levels of testosterone and other sex hormones. The levels are higher than in the embryo of any other mammal, higher even than in the blood of the wildest teenage boy. This shows that females in other species, including humans, have everything it takes genetically to be as aggressive as males. All they lack is the testosterone to flip the switches.

The easy explanation of the data from hyenas to hockey players is that testosterone causes aggression. But remember what happened when we assumed that serotonin caused aggression. Serotonin levels influence behavior, but the levels also are changed by behavior. The same is true for testosterone.

From song birds to squirrels, and mice to monkeys, an

aggressive encounter changes testosterone levels. Winners get a blast of testosterone; losers get a drain. Humans are the same. A group of varsity college wrestlers with roughly the same testosterone levels was checked before and after a competition. The winners got pumped; the losers went down. The same change in testosterone levels is seen in tennis players and even chess players.

Men don't have to be physical or even to beat someone in a competition to get a testosterone high. Brian Gladue and colleagues studied college students who had the opportunity to win or lose $5 on a coin toss. Even though the individuals had no impact on whether they won or lost, the winners were flushed with testosterone. Nor does real victory matter, only the perception of victory. Two men were seated across from one another at computer terminals. When their screens flashed the word "GO" they had to press a button as quickly as they could. After thirty tries the experimenter declared one of the students "the winner." The contest was fixed and the winner was picked by the experimenter; nevertheless, the "winners" showed an increase in testosterone levels compared with the "losers."

Just like with serotonin, the relationship between testosterone and aggression goes both ways: the human brain has been programmed to respond to testosterone with aggression and competitiveness, and to respond to competition and aggression by producing testosterone.

It's easy to imagine how this cycle could work in real life. Kids who are "naturally" high in testosterone are more likely to be aggressive and competitive, especially as they enter puberty and testosterone soars. Each victory (whether on the tennis court or the streets) gives them an additional burst of testosterone, which lifts their willingness to compete. But there is one thing testosterone—or any brain chemical or gene—cannot do, and that's to determine whether the kids get tough on a football team or in a gang, with a tennis racket or a gun.

THE MYSTERY OF
THE MURDEROUS MICE

Most of the time, science proceeds in small, not too unexpected steps. For example, since men were known to be more violent than women, and since testosterone was known to be essential for maleness, it was not that surprising to find a connection between testosterone levels and violence. But every once in a while, a result is completely unexpected. That certainly was the case in the discovery of the role of a unique gene in causing violent behavior in mice.

Not only was the discovery unexpected, so was the discoverer, Solomon Snyder. A professor of psychiatry at the Johns Hopkins School of Medicine, Snyder had never been particularly interested in studying aggression, but he had always been intensely interested in how the brain works. He started his professional life as a classical psychiatrist but quickly realized that understanding behavior—and ultimately treating psychiatric illness—would require understanding the brain at the molecular level. While doing an experiment to understand what happens to the brain during a stroke, he discovered the role of an important chemical—and its surprising side effect.

The chemical is called nitric oxide, and of all the neurotransmitters that brain cells use to communicate, it is surely the strangest. Unlike the other chemicals, it's a gas, the same gas that's found in car exhaust (although not the same as nitrous oxide, or laughing gas, used by dentists). The chemical is made by an enzyme called nitric oxide synthase, or NOS. Snyder is the person who discovered that one function of nitric oxide is to kill cells after a stroke, which makes strokes even more damaging.

This discovery led Snyder and his team to wonder how a brain would perform without the chemical. His theory was that recovery from strokes would be greater if the chemical were not present. It was a relatively simple matter to "knock out" the gene for the gas and breed a batch of mice that lacked any trace of nitric oxide in the brain. At first the NOS knockout mice seemed perfectly normal: they moved, ate, and groomed themselves just like any other mice. What a disappointment! Every scientist who discovers a gene hopes it's for something "important"; they don't become famous for discovering do-nothing genes. To be sure they weren't missing anything, Snyder and company cut open some of the mice brains. Again, nothing unusual. The only thing the autopsies did find was that, as expected, the knockout mice were more resistant to brain damage from strokes. This was good news but it didn't make sense. Why would mice—and humans—keep a brain chemical that's only function was to make strokes worse?

There was one unusual thing about the knockout mice. Every morning, the researchers would find one or two of them dead in the cages. They didn't appear sick, but they were dying at a very high rate. To find out what was happening, the mice were put under 24-hour surveillance. Video cameras discovered the secret: the males were killing each other.

When a normal mouse was introduced, the male knockout mice went crazy: kicking, biting, scratching, and shaking their tails. Compared with normal mice, the knockout mice attacked the visiting mouse four to five times more often. When a group of knockout mice were dumped together in a cage, the scene looked like a gang fight. Some of the mice got up on their hind feet like kangaroos to punch the other mice. None of the mice was willing to give up. A normal mouse will end a fight by rolling onto its back with its paws up, like a weaker dog submitting to a stronger one. But the knockout mice showed ten times fewer submissive postures. When one did try to give up, the other mice ignored the surrender and kept on fighting. The experiment had to be stopped after 15 minutes so the mice wouldn't kill one another.

The knockout mice weren't only interested in killing. They also wanted sex, lots of sex, and on their terms. When a normal male mouse meets a new female, his first reaction is to mount her. If she isn't in heat, she will make it known and be left alone. Not with the knockout mice. No matter how much the females screamed and fought, the mutant mice kept climbing on and trying to mount them, even when they weren't in heat. Their approaches came two to three times more often than normal.

The first theory for the bad behavior was testosterone, but their levels were normal. Maybe the mutants couldn't smell when the females were in heat? No, they could sniff out a hidden bit of cookie as well as any mice. The mutants also tested normal for strength and agility. They showed no signs of unusual bravery: when placed in an open field they were just as afraid as normal mice. They were just bad mice.

The researchers concluded that nitric oxide must function as a kind of brake on behavior. The knockout mice lacked the gene needed to make the chemical and didn't recognize the signals from males trying to surrender or from females declining sex. It isn't known yet whether nitric oxide plays a similar role in humans, or whether there are any naturally occurring mutations in the human NOS gene. What is clear is that the brain doesn't always work the way we think it does.

INMATE X

So far there is only one gene linked to violent behavior in humans. The clue that led to its discovery came from a man we'll call Inmate X, the single most important individual in the entire debate over the biology of crime. Inmate X was truly a prisoner of his genes, in fact, of just one gene.

Born in Holland, Inmate X was 23 when he was convicted of raping his sister. He was transferred to an institution for psychopaths, where he was described as quiet and easy to handle. In spite of this, fights occurred with other inmates. He was working in the fields one day and became upset when a supervisor told him to work harder. Inmate X stabbed the warden in the chest with a pitchfork.

A relative of Inmate X visited a doctor because she wanted to have children but was worried about her family history. She told the doctor that Inmate X wasn't the only one with a problem in her family. When she was younger, her own brother used to force her to undress at knife point. Another male relative had tried to run over his boss with a car. Two other male relatives had committed arson. Then there was a male relative who exposed himself, and another who was a voyeur. The problem of boys bothering their sisters was so common that some of the girls had fled their homes.

Her frightened inquiry led to a long scientific study of this single family. The bad behavior of the men varied widely, but there were common threads. All eight of the men closely studied were either mildly mentally retarded or borderline. The typical IQ score was 85. The only one to have completed primary school was also the only one with a regular job. All eight had repeated outbursts of aggressive, sometimes violent behavior. They were described as shy and withdrawn and often had no friends. A seemingly minor thing could set them off, and their response usually was way out of proportion to the provocation. They were especially likely to be aggressive during periods of one to three days when they slept little and complained of night terrors.

The problems with the woman's family stretched back for at least five generations. Thirty-five years before, one of her ancestors had done his own investigation. He visited all the men in the family and found that nine of them were retarded. Five additional cases emerged later. Inmate X himself had three affected brothers and two affected nephews, both of them related through sisters. Then there was his affected cousin, who was the son of his mother's sister. A maternal grandaunt's son was affected. So was a great-granduncle on the same side of the family. So were three sons of maternal great-grandaunts and two of their nephews through a sister. All together there were 14 men who showed the characteristic syndrome.

The family seemed to be suffering from a mysterious disease. The symptoms were retardation, aggression, and bad behavior. The victims were certain males; other men in the family were happy and healthy. Not one woman in the family had ever shown any sign of the disease.

It took a Dutch geneticist, Hans Brunner, to see the pattern. All the troubled males were related through their mothers. There was not a single instance of an affected father having an affected son. Whatever was making these men bad was passing through the females in the family, yet always sparing the women.

Brunner suspected the X chromosome, which is one of the two sex chromosomes. Males have an X chromosome and a Y chromosome, whereas females have two X chromosomes. Therefore a male always receives his X chromosome from his mother, never from his father. This means that traits that are controlled by genes on the X chromosome are always passed down to sons from their mothers, which was exactly the pattern seen in the family of Inmate X. Typically such traits are expressed more frequently in males than in females because males have no "good" copy of the gene to mask the "bad" copy, whereas females have an extra "good" copy of the gene on their other X chromosome. This is how many X-linked traits, such as color blindness and one type of hemophilia, are inherited.

Based on this clue, Brunner and colleagues set out on the arduous task of precisely mapping the gene, isolating it, and figuring out exactly what it did: what type of protein it made and how it functioned in the brain. They analyzed the DNA from family members to see if there were any genetic signposts, called markers, that were linked to the syndrome. The trial and error paid off when they discovered that all the problem men had one variation of a particular marker, and the healthy men had a different flavor. The responsible gene was none other than that for monoamine oxidase A, one of the enzymes that breaks down serotonin—already a suspect in violent behavior.

Brunner and company soon showed that Inmate X and four of his troubled male family members all had exactly the same mutation in the monoamine oxidase A gene: a single base change that told the cell machinery not to produce the enzyme. Twelve of the healthy men in the family did not have this mutation. The odds of that happening by chance were less than 1 in 1,000. The enzyme that was supposed to break down serotonin was dead in the violent men, and they were filled with a natural toxic waste.

The discovery of the "crime gene" generated headlines around the world and prompted criticism from scientists who refused to believe that something so simple—even simpler than the mechanism that determines blue eyes or brown eyes—could turn men into monsters. Doubters said the mutation alone could not cause such a personality change, that other genes must be involved, or it was something in their childhoods.

The doubt lessened when mice were engineered with the same mutation: they became crazed killers. Mice with a knockout for the monoamine oxidase A gene bit and attacked other mice without provocation. When they mated they squeezed their partners harder than other mice, and their partners squealed louder than other females. Soon the mutant mice were covered with scars from fighting and had red patches where the hair had been torn out by the roots.

Even this evidence wasn't enough for the critics, who continued to attack Brunner as a "biological determinist" and other nasty names. Sadly, the most important finding was largely ignored in all the commotion. Shortly after the discovery of the mutation in the Dutch family, scientists looked for the same change in a large number of DNA samples from a variety of people with mental problems, including violent criminals. Not one of them had the mutation, which is so rare that it's been called an "orphan mutation."

What this means is that the monoamine oxidase A mutation cannot be used to explain all human violence and crime, or even a substantial percentage of it. In fact, that was never the claim. What Brunner found was the source of abnormal behavior in just a single family, not in the population at large. What is important about the discovery is that it confirms the role of one particular aspect of brain chemistry—the monoamine system—in the biological part of aggression. It doesn't say anything about the environmental component, which takes a different kind of research.

ENVIRONMENT: THE FAMILY

Although genes and the brain chemicals they control clearly play a large role in anger and aggression, they are hardly the entire story. Even identical twins, who have exactly the same genes, can differ greatly in their levels of anger and aggression. Why?

Imagine the brain like a muscle. Genes determine its potential to be big and strong, but it's exercise that actually makes the muscle improve. Woody Allen will never have the same physique as Arnold Schwarzenegger, no matter how much he works out, because he doesn't have the necessary genetic makeup. But the reason Arnold is so huge isn't just because of his genes, it's also because of thousands of hours lifting weights.

The same principle applies to the brain circuits that determine anger and aggression. Genes play an important role in how easily and frequently a person experiences anger, but how the person actually behaves depends just as much on emotional and cognitive habits—habits learned by exercise and repetition. Learning to control and channel anger effectively is a lifelong task. It starts in childhood, and the first lessons come from parents and siblings. A baby watches an older sister stamp her feet and throw a tantrum. If the parents reward the girl by giving her what she wants, wouldn't it be logical for the baby to try the same tactic? More serious aggression also appears early, and much depends on how people react to the angry child.

Studies of twins suggest that the *general* parenting style, which both twins share, has little quantitative effect on their aggressive behavior. However, a new study of 708 families across the United States shows that children are sensitive enough to respond to *individual* differences in parenting. This study by researchers at George Washington University School of Medicine and Health, which included families with twins and/or at least two children of the same sex, found that if one twin is treated differently by the parents, even subtly, he or she will respond accordingly. The research shows that one of the most important things parents can do for their children is also the easiest: expressing love and affection. When parents tell their kids they love them, or hug and hold them, the child learns this is a healthy emotion. The lesson rubs off; children whose parents are warm and supportive get along better with siblings and other children. Parents also teach their children social responsibility by belonging to the PTA or encouraging the children to become involved in things outside school and the home. Not only is this logical, it is supported by the research at George Washington University, which shows a clear relationship between positive parenting and positive childhood behavior. Parents who really communicate with their children and enjoy doing things together are doing themselves and their children a favor.

Not surprisingly, dealing harshly or negatively with a child has the opposite effect of positive parenting. The more the treatment of the child is based on conflict and punishment, the more aggressive and antisocial the child will become. Negative parenting includes yelling, punishing, and rejection. Negative parenting also includes what appears to be the opposite of harsh treatment: ignoring the children. This is because the parent who says, "I give up, do whatever you want," surrenders to misbehaving and is sending the child a confusing mixed signal.

An important finding of the George Washington University study is that negative parenting has a statistically stronger effect—causing bad behavior—than positive parenting has in causing good behavior. Occasionally blowing up at a child, or even a well-deserved spanking, is not going to do permanent damage. The danger is slipping into negative habits, into a pattern of negative behavior toward the child, because bad parenting is what has the strongest impact.

Sadly, the parents who have by far the biggest influence on their children are the ones who beat and abuse them. The impact is well documented and undeniable, and always negative. The effects are apparent early on: kindergartners were twice as likely to be aggressive if they were being abused at home. And the damage lasts forever. Children who were identified by courts as having been abused or neglected were 42 percent more likely to be arrested for violence as adults than people who were raised without abuse.

The reason abused children don't treat others with kindness, even though they know what it's like to be hurt, is because they lack empathy. When a normal group of children aged one to three saw a child who was hurt or angry, most reacted with concern, sadness, or empathy. The abused kids reacted with fear or anger and even attacked the kids in distress. First the abused child tried to ignore the fussing child, and if the crying didn't stop, the abused kids punished the smaller child with physical violence. These aggressive children displayed the same behavior their parents showed to them—a pattern that would last a lifetime without intervention.

When Violence Makes Sense

Of all the things that determine whether people will be violent, aggressive, and antisocial, the most important is neither genes nor parenting style. The most important factor is not the type of brain a person has or whether they were abused as children. What matters most is geography.

The people in the inner cities are the same genetically as the people in rural areas, yet urban crime has spun out of control. We are the same genetically as the Nazis, yet we like to think of ourselves as more civilized. The level of crime in American cities, or the crimes against humanity in Bosnia, Rwanda, or Nazi Germany, can't be explained one bit by genetics, because genes don't change that fast. It's hard to say what is the cause of urban crime, or of genocidal atrocities, but it sure isn't biology.

Some people blame crime on race, which is genetic. Black men in America today are much more likely to be in jail than white men, and they are more likely to kill each other. They are not committing crimes because of their genes, however. Race in the United States actually works more like an environmental factor than a genetic factor. Being born black in America subjects a person to certain beliefs, expectations, and pressures that can lead toward aggression and crime. Being born into a black family means a higher risk of not having a father at home. These are not genetic or biological influences but environmental influences. In fact, the effects of race on crime rates disappear with intervening variables such as rural versus urban, rich versus poor, and employed versus unemployed. For example, the rate of violent crime in rural counties with large black populations is lower than the rate in white urban areas.

Although many studies have shown that individual differences in aggression and antisocial behavior are at least partially because of genes, it's important to remember that every one of these studies was done on a relatively homogenous population in a place like Iowa or Sweden. They were expressly designed to study individual differences, not group differences. The ideal experiment, which has never been done, would be to study a white baby from Greenwich, Connecticut, who was raised in the slums of Rio. Or a pair of identical twins born in Ethiopia who were separated at birth, one going to Israel and the other to Iceland. If such an experiment ever were done, the results could be predicted: social and cultural environment would play a very large part in predicting who commits crimes.

Moving a baby from affluence to the ghetto—any ghetto—doesn't change the baby's genes. The underlying flow of good or bad humors—of testosterone and serotonin and the mysterious workings of the physical brain—would not change. A happy baby is likely to remain a happy baby. An aggressive child will be aggressive in the suburbs or the slums.

Why, then, should we study the biology of aggression at all? Some critics have leapt to the conclusion that any genetic study of aggression must in fact be an attempt to prove that racial differences account for differences in crime rates. In short, that African Americans or whites (or whatever group you don't like) are genetically violent and predisposed to commit crimes.

The topic is so controversial that organizers were criticized for merely proposing a conference on the genetics of violence. When the meeting finally was held at an idyllic rural retreat in Maryland during 1995, the lectures were disrupted by angry protesters. Waving red flags, the demonstrators burst through an unlocked rear door and shouted slogans denouncing the conference as racist and genocidal. While some of the scientists sat there and took the abuse, a few stood up and shouted back. One distinguished scientist even got into a shoving match with a demonstrator. The outbreak of violence showed that the critics didn't understand genetics very well: they were confusing individual differences with group differences. It also showed the geneticists didn't understand human nature very well; that by emphasizing the role of genes, they could easily be criticized as downplaying the role of society.

Another perceived danger of genetic research on aggression is that it will be used to justify bad behavior. How can someone who is "sick" or "genetically motivated" be punished? It's easy to imagine how discovering the biological "cause" of aggression could lead to it being declared "normal" and therefore acceptable. If someone is naturally aggressive, then they should be free to act out their genetic destiny. What gives anyone the right to block the self-realization of others? Why shouldn't a man be able to get in touch with his inner gangster?

Defense lawyers have been trying this for years, routinely falling back on the insanity defense, which can be a biological excuse. The next logical step is the genetic defense. James Filiaggi of Cleveland was one of the first to try it.

Filiaggi went to his ex-wife's house and tried to enter. She called the police and begged them to hurry. While she was still on the telephone with the police dispatcher, James Filiaggi broke down the back door, chased her out of the house to a neighbor's, and shot her repeatedly until she was dead. Filiaggi's defense was that he "went haywire." His attorney, James Burge, argued: "Without proper medication, and a treatment protocol, he is absolutely not responsible for his acts." The defense said Filiaggi had a chemical imbalance in the brain that triggered uncontrollable rages.

The three judges who heard Filiaggi's neurobiology defense didn't buy it. He was sentenced to death. But lawyers are persistent, and they'll probably keep trying. As they do, judges and juries will need to listen closely to one of the main themes of this chapter: predisposition is not predestination. Genes and neurotransmitters and hormones may tip the scales, but people are not robots programmed by genes. There is plenty of room for free will and conscience in how we behave and in how we judge the behavior of others. Remember that simply having a Y chromosome, which makes a person a man, is already a strong genetic predisposition to crime. Should we then pardon a murderer simply because he is male?

The real reason to study the role of genes and biology in

aggression is to understand what can be changed and what cannot, what works and what doesn't. For example, the mistaken idea that expressing "justifiable" anger is useful, or even healthy, permeates our culture despite the fact that it only makes people more hostile. And the notion persists that crime can be reduced by filling more jails, despite the evidence that all this does is produce more criminals.

The reason for our success as a species is that we have made the most of our genes. We must not surrender to our harmful impulses, such as aggression, but rather channel them to our advantage—and for this, knowledge is the key. At the individual level, this means learning and practicing the emotional and cognitive habits that can turn the energy that we waste on anger to more productive endeavors. It means routinely counting to ten and "turning it over" before acting. It means using our brains to improve the world instead of hating or scapegoating any particular person or group. It means making character dominate temperament.

At the larger social level, it means reducing poverty, improving education, eradicating slums, eliminating racism, and instilling discipline and respect for self and others. Aggression can play a useful role in society when it is channeled into appropriate forms of competition such as sports, games, the market economy, or at times of national peril, into defense. The secret is not to deny aggression—it's not going away—but to make it work for us rather than against us.

FOUR

ADDICTION

Drinking, Smoking, and Drug Abuse

*Those who do not recover are people who cannot
or will not completely give themselves to this
simple program . . . They are not at fault;
they seem to have been born that way.*
—*BIG BOOK OF ALCOHOLICS ANONYMOUS*

Jeremy wakes up late. He can't find the alarm clock because
he threw it across the room when it went off the first time.
He realizes he's going to be late for work again, which has
been causing him a lot of problems; the boss is going to be
upset for sure. He regretfully thinks about what happened
last night. He was trying to be good, but then he ran into
some old friends and he got carried away and binged. When
he finally got home he was so wired he couldn't get to sleep
until 5 A.M. He promises to himself: never, never, never
again.

He showers but still feels groggy and sluggish. What the
heck, he thinks, as long as I'm this late another few minutes
won't make any difference. What he needs is some hair of the

dog. He pulls his stash out of the freezer. He grinds the drug into a fine powder and cooks up a double dose. The anticipation is delicious, and the first taste hits him with a jolt. Deep inside his brain, a minute trickle of dopamine is released, initiating a series of all-too-familiar chemical reactions. His heart rate and blood pressure rise almost instantly. As he begins to feel "normal," he thinks, "That's better, but that's it for today. Absolutely not one more."

At work, once the drug wears off, Jeremy finds it difficult to concentrate; he's irritable, edgy, not on his best form. Nevertheless, he manages to stick to his word and stay clean—for all of seven hours. On the way home he feels a headache coming on. A big headache. He remembers that there's a place he can score only two blocks from his downtown office. Within five minutes he's copped and used. It's not really a problem, he thinks as the drug kicks in, putting him back on track. "I can stop anytime I want to. I proved that today, didn't I? And besides, it could be a lot worse . . ."

Jeremy is an addict. He shows all the classic signs. He has increased tolerance; it takes a double dose to give the same effect that he used to get from a single dose. He is both psychologically and physically dependent; he feels abnormal and dysfunctional without the drug and exhibits withdrawal symptoms such as headaches. His drug use has become uncontrollable and is interfering with his work and his life; even though he wants to quit he can't. Instead, he tries to fool himself by denying that there's really a problem.

Jeremy's drug is caffeine.

His preferred dosage is a double espresso. His fellow users take it in machiatto, latte, cappuccino, a plain cup of Joe, Pepsi, Coke, tea, or chocolate bars. It makes no difference to the caffeine addict. Regardless of how the drug is taken, the effect is the same: a quick burst of energy, a rapid rise in respiration and heart rate, an increase in blood pressure, a pleasant lift. Caffeine does all this by reacting with receptors and enzymes for adenosine, a chemical involved in energy production. When caffeine enters the body for the first time it gives the user a mild buzz, but as the system becomes accustomed to the drug, regular users feel sluggish if they *don't* have it.

It might seem strange to use coffee to introduce the very serious, life-threatening problem of addiction. Caffeine is, after all, the most popular drug in the world—legally imbibed from Seattle to Singapore, the basis of whole economies, a staple of world trade. Unlike cocaine and heroin, coffee does not drive users to rob stores, kill, or prostitute themselves to support their habits. Unlike alcohol, caffeine does not physically eat away at the user. Nor does it kill with a slow lingering death the way cigarettes do. Coffee is legal, it's pleasant, and its side effects are mild.

But it's still a drug. A psychoactive drug. It produces altered physical and mental states and changes the way people feel. It has much the same addictive power as its more feared cousins, such as alcohol, nicotine, cocaine, and opiates. The definition of addiction is the compulsive use of *any* drug, despite adverse consequences, resulting in loss of control of intake. Some people are more susceptible to addiction in general, whatever the drug, because of common, genetically influenced personality factors and brain reward mechanisms. Not all addictions are exactly alike, however, and there are unique genetic, neurological, and social factors involved with different types of substances. In other words, being addicted to coffee is in some ways the same as being addicted to heroin, but in other ways completely different. The best way to devise strategies to avoid or stop substance abuse is to understand what factors are common to all addictions and what factors are unique.

THE "ADDICTIVE PERSONALITY"

The idea that different addictions tend to go together will come as no surprise to anyone who has attended a meeting of Alcoholics Anonymous. The air is heavy with smoke. Coffee is dispensed by the gallon. Cookies and other high-sugar snacks are consumed with abandon. And more and more, the members, especially the young ones, introduce themselves with, "Hi, my name is Dianne, and I'm an alcoholic and a drug addict."

Research bears out the relationship between the different forms of addiction. In study after study, abuse of alcohol, illegal drugs, cigarettes, and other substances are statistically related. For example, in one of our studies at the National Institutes of Health, more than half of diagnosed alcoholics were current cigarette smokers, representing a fourfold increase over those without an alcohol problem. Similarly, 70 percent of drug abusers had problems with alcohol, a much higher figure than in the general population.

What is the thread that binds these different forms of addiction? Many people recovering from alcohol and drug addiction believe they have an "addictive personality" that keeps cropping up even after they have quit using their drug of first choice. Heroin users who went cold turkey recall how they ended up substituting alcohol for heroin. Drinkers who go on the wagon may replace trips to the bar with trips to the shopping mall. Or a person who has gone clean of all substances may suddenly feel pulled toward promiscuous sex.

Despite such anecdotal evidence, many experts now are convinced that the so-called addictive personality is a myth. In one sense they are right: there is no *single* personality profile that describes all addicts. In fact, substance abuse usually has a bigger effect on personality than the type of personality has on substance abuse. However, there are certain core personality elements that are seen time and again in addiction. The key point in identifying these addictive personality traits is to recognize that addiction is not an event, it's a process with its own order and logic. Different personality factors play different, sometimes even opposing, roles in the stages of addiction.

The first step toward addiction is starting—picking up the first drink, lighting the first cigarette, or taking the first pill. The experts call this first essential step "initiation." It might seem obvious, but the person most prone to become an alcoholic—even a person who somehow had a 99 percent propensity toward alcoholism—would never have a problem if he never took that first drink.

The personality factor most likely to be involved in taking the first drug, drink, or smoke is novelty seeking. As discussed in Chapter Two, novelty seeking is a broad, genetically influenced temperamental trait that includes a desire for new sensations, dislike of monotony, and lack of inhibition. This is the perfect recipe for a first-time user of alcohol or drugs. She feels bored or unsatisfied, she wants to try something new, and she isn't afraid of the risk—whether it's the risk of getting caught, or getting sick, or getting hooked.

The relationship between novelty seeking and substance

abuse has been demonstrated in several different ways. Marvin Zuckerman studied college students in the 1960s, before illegal drug use was so common, and found that students who scored high for what he calls sensation seeking were more likely to try a variety of illegal substances. Psychiatrist Robert Cloninger studied a group based statistically on the population at large and found a positive relationship between novelty seeking and alcohol use and alcoholism. Interestingly, the relationship was strongest in young people aged 18 to 29 and tapered off as they reached 50 years of age and beyond. This suggests that when people are young, the desire to try something new, such as experimenting with alcohol, is a driving force, but as they get older other personality factors keep them abusing.

The second stage in addiction is continuing to use the substance, which is called "maintenance." Just because a person tries a drug doesn't mean he will continue to use it. For example, even though most American teenagers experiment with alcohol, tobacco, and, to an increasing extent, illegal drugs, only a fraction become heavy users. Most stop or learn to control their use. In fact, some products, such as cigarettes, are so noxious that novices have to force themselves to become regular users. Likewise, the first experimentation with alcohol often concludes with the tester passed out on the bathroom floor.

Why do people keep using substances that produce such strong reactions, are known to be unhealthy, and are, to varying degrees, socially discouraged? Two of the most important personality factors appear to be anxiety and depression. As described in Chapter Three, anxiety and depression are aspects of harm avoidance, a genetically influenced temperamental trait. High harm-avoidant people may use drugs and alcohol regularly to mask or "self-medicate" their negative feelings. If they discover that taking a sedative such as alcohol or barbiturates calms their nerves or that using a stimulant like cocaine or amphetamines lightens the blues, then they may choose to continue using those drugs. Ironically, they may keep using the substances even after the effect has worn thin, or to the point where the drug use itself becomes a major source of the anxiety and depression it once relieved.

Although several studies have found that substance abuse is related to harm avoidance and other measures of anxiety, depression, and neuroticism, the relationship is often complicated. For example, one large study of drinking behavior showed that harm avoidance was a major factor after age 50, but was relatively unimportant for people under age 30. Another factor to consider is that women tend on average to be more anxious and depressed than men, and harm avoidance seems to play a more significant role in their substance abuse. A third complication is that heavy drinking and drug abuse often *cause* depression, so long-term studies are necessary to determine which came first, the drink or the depression.

The final phase in the addictive cycle is stopping—or not stopping—which is called "cessation." Cessation means the deliberate, purposeful termination of an established habit or addiction, which is different than simply failing to maintain the habit or addiction. Deliberately breaking any habit is difficult. Only a small minority of alcoholics, drug addicts, and cigarette smokers who try to stop are able to remain clean.

The most important personality trait for quitting is likely to be self-directedness, which is a learned character trait rather than a core temperamental trait. Being self-directed means having a clear sense of purpose in life, developing good habits to act in accord with these goals, and being able to delay gratification. Self-directedness is related to the trait of conscientiousness, which describes the ability to manage desires and control impulses. The most important facet of conscientiousness for quitting is deliberation, the tendency to think carefully before acting.

Unfortunately, there have been relatively few large studies on the personality traits associated with successful recovery from substance abuse. Among our study participants at NIH who had quit drinking or cigarette smoking, we did see the expected results for self-directedness, conscientiousness, and deliberation. The interpretation is complicated, however, because these individuals were low on novelty seeking, which is one of the reasons for beginning substance abuse. There were several differences in other scales that made the results less than clear.

Perhaps the best evidence about the role of deliberation comes from talking to people who are recovering from substance abuse. They often stress how important it is to "think through" the drink or drug to avoid relapsing.

Ken, who's recovering from alcohol addiction, describes what happened to him on a business trip:

> There I was at 30,000 feet on my first trip out of town since I stopped drinking nine months ago. I had my sober friends' telephone numbers in my wallet and my AA literature in my briefcase. I was feeling pretty serene, pretty confident about not drinking.
>
> Then I heard that damn drink cart. There's something about the tinkle of ice cubes that . . . I don't know.

Maybe I really do have an alcoholic mind, because I suddenly wanted a drink so bad I could taste it. Hell, I could *feel* it.

I was just about to order a Jack Daniel's when I remembered something my sponsor told me to do when I was thinking about drinking: I remembered back to my last binge. The one that started out at the bar in the Ritz Carlton and ended up with me spending the night in the bathtub, naked, so it would be easier to wash off the vomit in the morning. I still remember the look on the face of the maid when she found me there passed out. Not a pretty sight.

So when that sky waitress asked what I'd like to drink, what came out of my mouth was "Diet Coke." Why it wasn't "Jack Daniel's" I'll never know. What I do know is I'm sure I've got another drunk in me, but I'm not sure I've got another recovery.

What kept Ken from drinking? It wasn't brain chemistry. After many years of alcohol abuse, his brain was telling him that a drink would feel good. And it wasn't likely to be genes or temperament. That's probably what started him drinking in the first place. What saved him was what he had learned, both from his Alcoholics Anonymous sponsor and from his own experience. This was explicit, conscious memory, not emotional memory. It was, in a word, character.

THIS IS YOUR BRAIN ON DRUGS

Kicking an addiction may be a matter of character or will power, but getting hooked in the first place is not. Addiction is not necessarily a sign of mental weakness, a lack of character, a symptom of psychopathology, or even a social disease. Rather, addiction is a disorder of the brain induced by chemicals that modify behavior. People become addicted to drugs for one simple reason: the drugs change their brains.

Perhaps the most convincing evidence about how addiction works comes from studying animals. If rats are given the choice of pushing a lever that delivers cocaine or another lever that delivers food and water, they start off pushing both levers about equally. But within a short time, they ignore the food and water and go straight for the cocaine every time. They will continue to do this until they die. Just like human addicts, they sacrifice their own welfare for the reward that cocaine gives their brain. Rats, mice, and other laboratory animals can also be taught to administer themselves amphetamines, morphine, alcohol, nicotine, and just about any other drug you can imagine. If they fail to get the drugs, they go into painful withdrawal, so they keep on pushing the lever—until they die.

Although individual drugs work differently on the brain, the major addictive substances share one mechanism: they activate the nucleus accumbens, the brain's reward center. This is the part of the brain that recognizes a new drug and says, "Hey, I like that." When the drug stops appearing, it says, "Something's wrong here." When drug-addicted animals were tested with PET scans, their brains showed a hot spot of metabolic activity directly over the nucleus accumbens. The hottest area of all was the outer shell of the nucleus, a region that connects directly to the limbic region, the seat of emotions. The drugs were acting on the connector between midbrain and forebrain, the perfect place to link good feelings to specific behavior. This is where the brain says, "If it feels good, do it."

To figure out what cocaine was doing in rat brains, miniature probes were inserted into the shell of the nucleus accumbens and samples were taken before, during, and after drug use. Of the many signal molecules present in the brain sap, only one was specifically elevated: dopamine, the "pleasure chemical." Moreover, when the dopamine fibers were intentionally damaged in addicted rats, they stopped taking drugs. This showed that it was the brain's own dopamine—activated by the cocaine—that was stimulating the rats.

The addicted rats illustrate how pointless it is to argue about whether drug addiction is environmental or genetic, a social disease or a medical problem. Obviously, their drug use was 100 percent environmental: cocaine is not produced by the body; it has to be ingested. If the rats hadn't been exposed by drug-pushing scientists, they never would have become hooked. On the other hand, once the rats had their first taste of cocaine, they became physiologically and biologically driven to keep using it. The need for cocaine became as strong, and finally stronger, than the need for food. It wasn't the scientists who kept the rats hooked; the rats were betrayed by their own brains. The rats didn't live in especially bad neighborhoods, they weren't abused as babies, they weren't members of an oppressed minority, they weren't depressed about being unemployed. The rats became drug addicts because the drugs rewired their brains.

Human brains work the same way. Anna Rose Childress and colleagues at the University of Pennsylvania studied cocaine addicts who were seeking help. When the addicts were allowed to hold crack pipes or were shown videos of people doing cocaine, their brains lit up under PET scans.

The hot region was the mesolimbic dopamine area from the prefrontal area down through the amygdala. The actual nucleus accumbens is too small to be resolved in PET scans, but is known to be part of this circuit. Just the idea of drugs was enough to excite their cocaine-wired brains.

After the addicts had been off the drug for about a month, the pattern changed. The dopamine circuit turned ice cold. There was hardly any activity there, and the cells themselves appeared to be damaged as if from a head injury. The cells had been flooded for so long with so much dopamine that they had lost their normal responsiveness. What that means is that even though the dopamine level in the brain was "normal," the person would feel hungry for dopamine because the brain had become so desensitized. It would take higher than normal levels of dopamine to produce normal feelings, which is why people crave drugs. The brain gets to the point where it only feels normal when it's chemically altered.

Fortunately, the changes that drugs cause in the brain are usually not permanent. They can be reversed once the drug use is stopped. PET scans of recovering cocaine addicts show that the dopamine circuit is pretty much back to normal after one year of abstinence. The brain has a certain plasticity in this regard, which means that it is possible to recover from addiction if the brain is denied drugs. The first weeks are the most difficult because the brain feels deprived, but gradually the chemical balance will return to natural levels. Just as drugs can rewire a normal brain into an addicted brain, cutting off the drugs can return the brain to its original state, or close to it.

ADDICTION GENES

Any brain can become addicted, but some brains are easy targets, and the reason is largely genetic. Studies of animals have shown that virtually every possible reaction to every different drug has some degree of genetic control. Strains of mice, for example, can be genetically engineered to prefer alcohol to water. Or mice can be bred so that the slightest amount of alcohol makes them stumble and pass out, while others can be bred to drink much more without getting tipsy. Some strains of mice react to alcohol withdrawal with seizures, while others have no trouble going on the wagon. Mice also can be bred to prefer their drinking water laced with morphine. For some strains, morphine works as a powerful analgesic, allowing them to walk on a hot plate without flinching. Other strains get less relief from the drug. Caffeine makes some mice more jumpy and hyperactive than others. For some inbred animals, a tiny dose of diet pills is enough to make them stop eating while others need bigger dosages. There are mice that react to cocaine with wild, frenzied activity, and some mice that remain calm and quiet. Nicotine makes some strains of mice learn better than others, while it makes still other strains go into seizures.

All these differences are purely genetic. The mice were tested in the same environments and exposed to the same influences; the only difference was breeding. This does not mean there is one single "addiction gene." In fact, different reactions are controlled by many different genes, and the genes that control reactions to one substance are only partially the same as the genes that control reactions to other substances. Genes don't control addiction; rather they control an animal's reaction to a substance, how much of the substance the animal can tolerate, and how strong an influence it has on behavior.

For humans, just like for animals, nearly every aspect of every type of substance that can be abused—from alcohol to amphetamines to nicotine—has been shown to have genetic influences. Genes can influence not just whether a person uses a drug but also how the drug affects the person. For example, amphetamines act as stimulants for most people, but for children with attention deficit hyperactivity disorder, a largely inherited condition, the same drug actually reduces activity and improves attention. There is little evidence yet linking specific illegal drugs to specific genes, and only a modest amount of research has been done on the genetics of smoking. So far, the big research money in this regard has been spent on alcoholism.

ALCOHOLISM

It's not surprising that alcohol is the drug that has been studied most at the genetic level. Alcohol is the drug humans have abused for the longest time, and its health and social effects are enormous. Everyone has a story about an alcoholic who grew up in an alcoholic home, and experience shows that alcoholism runs in families. In fact, one of the best ways to predict whether someone is going to be an alcoholic someday is to look at his closest relatives. Five times the normal number of the male relatives of alcoholics are themselves alcoholic.

However, just because something runs in families doesn't mean it's genetic. Maybe children drink because they are imitating their parents, or because life is miserable in an alcoholic home. To test this, scientists looked at alcoholic fathers who put their sons up for adoption. Both in the United States and in Sweden, the biological children of alcoholics had a four times higher risk of becoming alcoholics themselves than the children of nonalcoholics, even though they'd been separated from their alcoholic parents after the first few weeks of life. When the experiment was reversed— when the children of nonalcoholics were adopted by alcoholics—the children had no increased risk for alcoholism, even though they were exposed to it in the home. Likewise, when one son of an alcoholic was raised by a different family, he had the same rate of alcoholism as his brothers raised by the alcoholic father. Apparently just being exposed to alcoholism at home is not enough to turn a person into an alcohol abuser. Some children of alcoholics might even try harder than others to avoid alcohol, since they know first hand the damage it can do.

Genes clearly have a strong influence on alcoholism, but not every alcoholic is equally influenced by genes. Alcoholics are not all the same, nor are their reasons for drinking. Cloninger has divided alcoholics into two groups. Type II alcoholics are almost exclusively males who start drinking when they are young and often exhibit antisocial behavior. This is the bingeing, bottle-throwing, door-kicking drunk. Type I alcoholics are just as likely to be women as men, they typically develop a problem during middle age, and they are steady drinkers. This is the person who seems to be nursing a beer all day long. Adoption studies in Sweden showed that Type II alcoholism is very strongly genetic, while Type I alcoholism, especially in women, results more from environmental effects.

Even among "moderate" drinkers, genes may play a significant role in alcohol consumption. For example, in a Swedish study the correlation of how much twins drank was twice as high for identical twins as for fraternal twins, which shows a strong influence of heredity but no obvious effect of upbringing. In fact, the only dimension of alcohol consumption that is strongly influenced by family environment is teenage abstinence—probably because drinking is prohibited by the parents.

The twin studies show the genes have a controlling influence on alcoholism, but they don't show how. One likely possibility is that they influence tolerance to alcohol. For a relatively small number of people—estimates range from 3 to 20 percent of the population—alcohol is a deadly poison. If they drink, these people risk being destroyed mentally and physically. For them, alcohol is a killer, whether directly through acute alcohol poisoning, liver disease, or heart failure, or indirectly through malnutrition, a car accident, or suicide. They drink not because they want to but because they cannot stop. They are alcoholics.

What makes an alcoholic? The answer is complicated, but part of the reason is simple metabolism. Even before they've ever had a drink, some people will react differently to alcohol. For lack of a better term, we call them prealcoholics. Because of genetically determined biochemical pathways, the bodies and brains of prealcoholics react to alcohol by wanting more. People who aren't prealcoholics can drink much less before they become sick and must stop. Most people have a biological brake that stops them from drinking to excess, at least most of the time. Prealcoholics, because of a genetic difference, lack this brake. In the worst instance, the controls are actually reversed: the brake has been replaced by an accelerator, and while a normal person slows down after a few drinks, the prealcoholic is still picking up speed.

Finding the genes that make prealcoholics is now a major goal of the National Institute on Alcohol Abuse and Alcoholism, but despite extensive and costly research the results are still slim. For example, in 1990 Kenneth Blum and his colleagues at the University of Texas Health Science Center in San Antonio described a study in which they compared 35 alcoholic patients and 35 others and found one gene, which makes a dopamine receptor called D2DR, that was different. However, when other scientists tried to replicate the finding, they failed. Then it was discovered that the type of gene varied among ethnic groups, which might have thrown off the results. The killing blow to the "discovery" came when it was shown that the DNA sequence was in a part of the chromosome not known to have any functional significance. In other words, they had discovered a gene that didn't do anything. Perhaps a genomewide scan will give more reliable results, which is the current strategy of a nationwide consortium of alcoholism investigators and geneticists.

Although progress on finding alcoholism genes in humans has been slow, the research in laboratory animals is more successful. By comparing genetically pure strains of animals, it's possible to see which genes are involved in the body's response to alcohol. Perhaps the most significant advance, however, came from an unexpected source: the study of aggression. Scientists bred a strain of mice that lacked the gene for one of the serotonin receptors, called the 5-HT1B receptor, to test the idea that serotonin is involved in aggression. It was: the mutant mice were much more aggressive than normal. Since aggression and alcohol often go hand-in-hand in humans, the scientists wanted to see if these mean mice might also have a drinking problem.

The mice were given a choice of water or water spiked with various amounts of alcohol. Although both the mutant and normal mice consumed some alcohol, the mean mutants found it much more difficult to stop. They drank twice as much alcohol as normal mice. They swilled away a solution with 20 percent alcohol, or 40 proof, the same as a strong highball. Since high initial tolerance to alcohol is a characteristic of people who become alcoholics, the mutant mice were tested for their sensitivity to alcohol by putting them on a grid to see if they walked a straight line or stumbled about. The mutant mice were much less affected by the alcohol than were the normal mice. Just like human prealcoholics, the mutant mice could really hold their liquor—and kept on taking more.

These mice confirmed one of the most important findings about human alcoholism, the link between initial tolerance and alcohol consumption, and amazingly showed that just a single gene could affect both processes. Whether a single gene has such a role in humans is still not known.

Ethan's Tale

Our British friend Ethan is at the far end of middle age, slim, handsome, and articulate. With his clipped accent, Oxford education, and fine tailor, he could be mistaken for a character actor for the BBC. His weakness, however, is drink.

"I started drinking when I went away to university at age 18," Ethan said. "Right from the beginning I was the one who could really hold his liquor. When everybody else was plastered to the gills I'd be right as rain. I won a lot of money at late night poker games that way. They'd barely be able to see their cards, much less hold them, but I was sharp as a tack."

Ethan's high tolerance for alcohol probably existed even before he started drinking and further increased the more he drank. This preexisting tolerance has been found in studies of the young sons of alcoholics. When boys were given small doses of alcohol, the sons of alcoholics were far better able to hold their liquor than other boys. For example, when they were measured for how well they could stand straight, the sons of nonalcoholics swayed four times as much as the sons of alcoholics. The sons of alcoholics also were less likely to feel nausea, dizziness, or drunk. The alcohol even had different effects on hormone production and brain waves.

Ethan recalls: "For about seven years I had a good time drinking. People loved having me to parties. I was known as somebody who was witty, lively, a wonderful conversationalist. Of course part of the reason I was so bubbly was that I'd usually consumed a pint of gin before I even arrived at the party."

This early stage of alcoholism is called the adaptive phase. The alcoholic's metabolism is changing to allow the absorption of more and more alcohol, and his central nervous system is altering to become increasingly dependent on alcohol's stimulatory properties. His liver, the main site of alcohol detoxification, increases its output of the microsomal ethanol oxidizing enzymes that convert alcohol to less harmful byproducts. The mitochondria, the cell's power stations, become enlarged and misshapen to accommodate an increasing proportion of fuel being provided in the form of alcohol.

In the brain, the cell membranes are changing. The

membranes are made of a fatty material that is especially susceptible to the solventlike properties of alcohol. Low doses of alcohol activate the cell membranes leading to the release of chemicals that contribute to euphoric, pleasurable sensations. Higher doses decrease membrane activity, leading to the slurring of words and dizziness. The brain cell membranes in an alcoholic respond to the flood of alcohol by becoming tougher. They change their shape and composition so that more and more alcohol is required to produce any effect. To get any pleasure, the alcoholic must drink so much that the body and brain's defense mechanisms are constantly on red alert.

This ability to adapt is what makes alcoholics unique. Because of their genetic makeup, their bodies and brains respond to alcohol with increased physical and mental tolerance. Nonalcoholics do not have the genetic makeup to withstand this type of battering. For them, drinking this much would make them ill or pass out, and they would suffer a wicked hangover. They would be physically incapable of consuming as much alcohol as someone like Ethan. But even a hardened alcoholic can't drink forever.

"Once I discovered the Beefeater martini with three olives—a drink we don't have in England—it was basically straight to hell in a handbasket," Ethan says.

At this point, Ethan entered the second or middle stage of alcoholism, which is marked by loss of control, unpredictable tolerance, and the beginnings of physical dependence. He would earnestly promise himself to limit his drinks at a business lunch or social event, only to find himself going far beyond what he'd planned. And he could no longer be exactly sure what effect a drink would have. Sometimes it would take five or six martinis to feel anything, but other times a single glass of wine would make him woozy and furry mouthed. He went from taking a drink in the morning to cure a hangover to taking a drink in the morning to cure the shakes. His work suffered, and after many warnings he was fired from his bank job.

"It wasn't easy finding a new job," says Ethan. "Nobody was buying my story about leaving the business because of my egalitarian, progressive social ideas. They could smell the reason why. I always took my flask with me to the interviews."

The gradual switch from wanting a drink to needing a drink is a consequence of the brain's remarkable ability to adapt to altered circumstances, a process called neuroadaptation. The brain struggles to remain in balance. When alcohol is present, the brain responds by trying to restore the balance. The major changes are in the dopamine circuit in the nucleus accumbens—the same circuit involved with cocaine—and in the system that regulates the firing of nerve cells throughout the brain.

Alcohol, just like cocaine, releases dopamine in the nucleus accumbens, leading to a sense of reward. But as the alcoholic drinks more, the dopamine cells wither and shrink, so when alcohol is not available the person feels lousy—and wants another drink.

The other effect is more generalized. GABA, or gamma-aminobutyric acid, is an amino acid that inhibits neurotransmission throughout the brain. The receptors for GABA are especially sensitive to alcohol, and seem to be responsible for many of its pleasurable effects, as well as the unpleasant effects of stopping. Chronic alcohol consumption alters the GABA receptors so that when alcohol is not available they can't inhibit brain signaling, making the person susceptible to strong feelings of anxiety and to seizures. When mice were bred with a faulty GABA system, they were less affected by alcohol. Although it's not yet proven, genetic differences in the GABA system could very well explain different reactions to alcohol.

Out of work and on his own, Ethan progressed rapidly to the final stage of alcoholism. "I was very proud of the method I devised to get gin inside me in the morning," he says. "No matter how much I drank in the evening, I was careful to leave a little extra for the next day. The problem was drinking it. My shakes were so bad that I couldn't drink it from a glass or even straight from the bottle; I would have cracked a tooth. So, I'd leave out a mixing bowl, the largest I had, and pour the gin in with one hand and the tonic with another. And then I'd drink it with a straw. That I could manage. The sad thing is that I was convinced that this proved that I was not an alcoholic. After all, I told myself, I was using my mental facilities to solve a difficult problem. Yes indeed, I was very intelligent when it came to drinking."

In the final stage, the tolerance to alcohol is lessening because of the damage to the liver and brain. The brain cannot snap back from the punishment and now requires alcohol to feel in balance. The withdrawal symptoms worsen to the point where the alcoholic spends most of his time drinking because to stop would be excruciatingly painful. The result is a body and brain that exists only to drink.

Ethan was lucky. He stopped drinking 12 years ago, his health recovered, and he returned to his old job. His career did not advance as far as it might have because of the missing years, and he's still paying for the enemies he made when he was too drunk to care, but he is alive. He now has an excellent incentive to stay sober: his son is an alcoholic. The son got into AA at age 22, and with his father's encouragement, is currently sober.

Plugging the Jug

Ethan didn't need a neuroscientist to tell him that his dopamine and GABA circuits had been altered by alcohol; his addiction was painfully obvious every day of his life. Nor did he need a geneticist to explain that his condition might be inherited; his own son was proof enough that alcoholism could be handed down from generation to generation. But despite the neurochemical and genetic odds against him, Ethan was able to get sober. How?

"It was simple," says Ethan. "I stopped drinking." His words, overly simplistic as they might seem, reflect the one universally accepted principle of alcoholism treatment: the only consistently successful treatment is total abstinence. Not cutting back, not switching from hard liquor to beer, not drinking only after 5 P.M.; just plain putting the "plug in the jug." Alcoholics who have tried to "control" their drinking have learned, often the hard way, that it just doesn't work. This is why epidemiological research shows that the best predictor of relapse into alcoholism is previous relapse.

Often alcoholics make the mistake of thinking that their drinking is a result, rather than a cause, of emotional difficulties and seek help from psychiatrists or psychologists. Although this is one route into treatment, there are many mental health professionals who are still unaware of the signs of alcoholism and attempt to treat their patients with yet more drugs, including sedatives, which only worsens the problem. If you think you have a problem with alcohol, you do, and more likely than not the problem is partially genetic.

EXQUISITE NICOTINE

Oscar Wilde quipped that, "A cigarette is the perfect type of perfect pleasure. It is exquisite, and leaves one unsatisfied. What more could one ask?" Sartre put it more existentially: "A life without smoking [is] not worth living."

Along with caffeine, nicotine is the world's most widely consumed psychoactive drug. The numbers are staggering. More than one quarter of the world's population, some one billion people, consume nicotine in cigarettes, cigars, pipes, snuff, and chewing tobacco. All this tobacco use comes at a high price. Each year, approximately 400,000 deaths in the United States are attributed to cigarette smoking, at a cost in medical expenditures of $50 billion. The main killers are lung cancer, emphysema, heart disease, arteriosclerosis, and stroke. Smoking-related illness accounts for more than one-quarter of all deaths for people 35 to 64 years old, and between one-third to one-half of regular smokers will die of their addiction. Each cigarette robs a regular smoker of 5.5 minutes of expected life.

Why do people continue to smoke when it can kill them? It's not that they don't know better. Anyone who has hacked and coughed through the very first cigarette knows that smoke is not a benign substance. In fact, the bad effects of the "noxious fumes" on respiration have been recognized ever since the introduction of tobacco from the New to the Old World, and by the 1700s whole books were being written about tobacco's effects on cardiovascular function. Since then, public education campaigns, legislation, and social custom have ensured that the dangers of cigarette smoking are known to all. Yet people continue to light up.

The reason, we were told in 1996 by the Food and Drug

Administration, is that nicotine is addictive. This shocking "discovery" was challenged by the tobacco industry and others who profit from addiction. The truth has been known for centuries, however, and for evidence you need only ask any regular smoker who has run out of cigarettes and picked butts out of the ashtray and hungrily smoked them to the finger-burning nub. Or visit the smoking area outside any hospital and watch the doctors and nurses getting their fixes.

While we have known for some time that nicotine is addictive, what we are just now learning is how the drug works on the brain. The latest research shows that nicotine acts through two distinct brain mechanisms. Each one alone would probably be enough to make nicotine a habit-forming drug, but acting together, they weave a fine, tight web of addiction.

The first mechanism comes into play when nicotine reaches the brain, which takes only a few seconds when it's ingested in smoke. The drug binds to a special type of protein called neuronal nicotinic acetylcholine receptors, which are involved in learning and memory. Normally the receptors are activated by acetylcholine, a natural human substance, but for some unknown reason they can also be triggered by nicotine, a plant poison. This is what makes nicotine a brain stimulant, how it "perks up" a user and makes concentration easier. One of the questionnaire items most often endorsed by smokers participating in our research studies at the NIH is "I smoke when I need to concentrate," and experiments have shown that nicotine does indeed improve short-term learning and memory tasks. For a while, nicotine was even used to treat Alzheimer's patients because it seemed to make their brains function better.

The second mechanism occurs when nicotine stimulates

the release of dopamine in the nucleus accumbens, the exact same mechanism used by cocaine, amphetamines, and morphine. In fact, if you monitor the brain scans of rats injected with drugs, it's impossible to tell if the drug is nicotine or cocaine.

In addition to nicotine, cigarette smoke contains substances that inhibit the brain enzyme monoamine oxidase B, or MAO-B. Heavy smokers have 40 percent lower levels of MAO-B in the brain than other people. MAO-B's function is to degrade the monoamines—dopamine, serotonin, and norepinephrine—which play such a central role in how people feel. In fact, monoamine oxidase inhibitors are used as antidepressant drugs, which means nicotine may actually make some people feel less depressed. Studies show that people who say they have more than the "normal" amount of depression are four times more likely to be smokers than other people.

So, cigarettes make you feel smarter, better, happier. No wonder people don't stop.

Smoking Genes

When people hear about research on the role of genetics in smoking, their usual reaction is a puzzled frown, or a sneer, followed by, "That can't be genetic. I used to smoke a pack a day but I quit." The reason for the skepticism has nothing to do with science and everything to do with perception and politics. Alcoholism has been regarded as a disease for more than 50 years, and people are comfortable with the idea that diseases can be genetic. But until recently, smoking was thought of simply as a "habit," which most people don't associate with genes. This, together with the strength of the tobacco lobby, is why there is a whole branch of the National Institutes of Health investigating drinking—the Institute on Alcohol Abuse and Alcoholism—but only a handful of labs around the world looking at the genetics of smoking.

The first twin study of smoking goes back to 1958, when R. A. Fisher noted that concordance rates for smoking were higher for identical twins than fraternal twins, a key indicator pointing at genes. Since then, there have been some 18 different studies that found identical twins more likely to both smoke or both not smoke than fraternal twins. Heritability estimates range from 28 to 84 percent, and a summary of the results shows a mean heritability of 53 percent. The twin studies show no evidence that family environment has any effect on smoking; the increased smoking by the children of smokers is because the parents are passing on their genes, not their cigarettes. In our society, all children are exposed to smoking, but in terms of "environmental" impact on whether they start smoking, peers probably have more influence than parents.

The twin studies then were expanded to see what happened if one twin quit or tried to quit. Would the other twin have the same success or failure? The results showed that genes influenced whether a person started to smoke and also the likelihood of quitting, but the genes are different.

The genes themselves have not been identified. A good place to start looking would be the genes linked to novelty seeking, especially related to dopamine signaling. Recently our lab and other researchers at NIH have found an association between dopamine genes and the ability to quit smoking. If this finding holds up, it could offer a whole new approach to smoking cessation. Genes involved in continuing to smoke might be related to genes for harm avoidance, especially serotonin. One area that is completely unexplored is genes that control sensitivity to nicotine, perhaps the genes for the nicotine receptors themselves. Mouse studies have shown that genes have a major role in the initial stimulation by nicotine, which is a good predictor of subsequent tolerance and addiction.

Quitting Tobacco

The best advice about cigarette smoking is the simplest: don't start. Nicotine is a powerfully addictive drug. Most people who smoke on a regular basis do become addicted. This means that nicotine works differently than alcohol, because many people drink on a regular basis without becoming alcoholics.

Quitting is difficult because of the three unique brain mechanisms that hook and reward the smoker, plus the genetic role, and the strong psychological and behavioral reinforcement of the smoking habit. Even though more than 80 percent of smokers say they want to quit, only about 7 percent are successful in any given year. The relapse rate for smoking is as high as it is for heroin use.

But difficult is not impossible. The good news is that many good strategies for quitting have been devised. If one method doesn't work, try another. Be prepared to fail the first time you try to quit. Even if you just quit for a few days, that's a plus because every day without smoking is one more day that you aren't damaging your body. Remember, it's the total time that you smoke that affects your life span, not how many times you stop and start.

Thomas, one of our research subjects, started smoking as a teenager, and by the time he was 47 years old, he smoked three packs a day. When asked how he managed to smoke that many cigarettes in a day—about four an hour—he said he had ashtrays in every room in the house, and frequently he'd have one cigarette going in the kitchen, another in the bathroom, and another in the living room. "Sometimes I'd forget I had a cigarette going in the ashtray right in front of me and light up another," he said. Thomas also was a recovering alcoholic, with a family history of addiction.

His dentist noted a small white area on the inside of his cheek. He went to an oral oncologist for a biopsy. While he was waiting for the biopsy test results, which took a week, he vowed to give up smoking on the day he got the results. He asked the doctor about quitting and the doctor prescribed chewing gum laced with nicotine.

The biopsy results came back positive. Thomas had extensive surgery, including the removal of part of his tongue, which made speech difficult. Even with such a strong incentive, Thomas still had trouble quitting. The first time he tried the nicotine gum, he couldn't stand the taste, so he quit smoking cold turkey. After a month without smoking, he found the unused pack of nicotine gum. "It cost 30 dollars," he said. "So I figured I might as well give it one more try."

He still didn't like the taste, but he loved that old nicotine lift. "Pretty soon I just *had* to have that gum. When I got up in the morning, after a meal, watching TV—basically any time I would have smoked a cigarette. I knew what was happening, but I just couldn't resist the cravings." Pretty soon, Thomas was back on nicotine. The delivery system had changed but not the addiction. It took another three months and a doctor's help to quit again, this time for good.

Thomas's story illustrates three points: nicotine is strongly addictive, no matter how it's delivered. Addictions go together: Thomas's smoking seemed linked to his alcoholism and his family history of addictions. The third point is that it's possible even for such a heavily nicotine-dependent person with a genetically addictive personality to stop smoking.

THE ENDURING ATTRACTION

There is a strong human desire to experience altered states of consciousness, to change the brain reactions. From the toddler who likes to spin until he's dizzy to the father who pours a relaxing drink after work, the desire to alter the chemical mix of the brain is powerful. Perhaps this is a genetic impulse common to the species. Some cultures use drugs for religious ecstasy, others to relieve social tension. Some individuals try drugs to open their minds, others to close them. Some people want to speed up, others to calm down. People want to change the way they feel for different reasons, but the techniques are the same: slow down a synapse there, increase a neurotransmitter here, turn down this gene, turn up that one.

While some scientists are focused on the evils of alcohol, tobacco, and dangerous drugs, others are cooking up new batches of "designer drugs," emotions in a bottle. If history is a guide, we will keep experimenting with altered states of consciousness whether they come from meditation or prayer or chemicals.

FIVE

SEX

Men and Women in Love

We are never so defenseless against suffering
as when we love.
—SIGMUND FREUD

Paul and Madeleine met briefly when they were in college. It was during summer vacation at a northern lake, a hot night in a crowded bar. They clicked right away, both feeling they were finding someone they had known for years. With a few words and gestures, not sexual but warmly sensual and full of promise, Madeleine slipped into his heart like a shiny blade. Paul was not just another object of her flirtation, though; he seemed special, perhaps the one man who could understand and love her complicated, troubled soul.

But she was off to Paris, he to a career in the city. They wrote occasionally but didn't see each other for seven years, during which time they pursued other loves, conquests, and affairs with varying degrees of seriousness and little commitment. In the back of their minds, although they never voiced this hope, they nourished the thought that some day they would be together. Rather than keep them apart, the distance made their relationship grow in their imaginations, each projecting dreams that the other would realize. Paul often thought of the cool taste of her lips, and the way their bodies fit together. Madeleine remembered his indulgent smile and the shelter he built for her in his arms.

Finally, after seven years, they set a date to meet again. The risk was great because expectations had grown wildly, with no check from reality. Wasn't it better just to keep writing letters, to keep working on this perfect relationship? What if they woke up together and found that they didn't even like each other?

Paul was so excited he couldn't sleep on the long flight to the island. Madeleine was there already, waiting anxiously in a tiny bungalow, where everything was moist from the sea. When he arrived, they embraced and fell onto the bed. The sound of waves washed over them, and the spinning ceiling fan clicked gently above their bodies.

They spent a week together, sunburned and peaceful, more relaxed than they could imagine. Their reunion was everything they dreamed it would be: passionate, real, their connection absolute. When the week ended, Paul and Madeleine parted easily knowing that their fate was sealed; it was only a question of time before they would be together. Their questions had been answered, and they would pair for life.

While Paul went back to work, Madeleine returned home leisurely by ship. The ship's captain was strong and handsome, fascinating and free. Madeleine felt inexplicably and hopelessly attracted to him, but she knew that if she indulged herself it would be impossible to return to Paul. Madeleine saw herself swimming toward the captain, and when she hesitated for one second, he took her hand and tugged gently, bringing her in as easily as he would land a big fish for his dinner.

She called Paul long distance to tell him the story, in excruciating detail, and he felt revulsion. He tried to turn the pain into anger at her, or at the captain, but he couldn't. Madeleine said she was confused and didn't know what to do. She was torn, she told Paul, and didn't love him any less.

When she called again a few weeks later, Paul had been drinking and a woman he barely knew was in his bed. Madeleine's first words were, "You're not alone, are you?" He answered truthfully, "No." Madeleine was furious. She ripped the phone from the wall and smashed it onto the floor. She didn't need this kind of pain, not now when she was so confused. In her mind, Paul had destroyed their relationship. Of course, her affair with the captain never amounted to anything; he was married and had three kids. Madeleine's feelings toward Paul simmered for years until she convinced herself that she hated him. For quite a while, Paul thought of her every day, but eventually she drifted free of his memory. They never exchanged another word.

What happened to Paul and Madeleine? On such a huge planet and in time infinite, how did the paths of two people come together and become one? What force threw them into each other's arms with such passion, and then propelled them apart so violently? The answer is simple and complicated; the force is greater than anything known to our kind: Love. Falling in love—or falling out of love—is among the most memorable emotional experiences most people will have during their lifetimes. The choices we make about love are among the most important we are allowed, both in terms of happi-

ness and fulfillment, and of passing on the essence of ourselves, our genes.

The reason these emotions are so strong and memorable
is that romantic love is, before anything else, an expression of
sex and reproduction, and sex and reproduction are among
the most powerful of all human drives. Human love cannot
be reduced to mating, however, and perhaps more than any
other emotion it is influenced by chance, culture, social custom, serendipity, even "the stars," which explains the endless
variety of stories, songs, poems, and movies about love. Bur-
ied under the mountains of words we have invented to explain and glorify romantic love, its fundamental core has always been the physical union of a man and a woman. The
need for this union is among the most deeply rooted, innate,
and genetically programmed of all human behaviors.

The power of love is simple. The way humans pass on
our genes is through sexual reproduction. People with genes
that somehow made them incapable or disinterested in having sex never had children and therefore didn't pass on those
genes. They may have been wonderful, even extraordinary
people. All their other genes may have been great—genes
that made them kind or generous or strong or brilliant or
beautiful—but if they didn't have sex, those genes were lost
forever. People with genes that inspired sex and attracted a
partner had children. The children had children of their own
and so on down to us. Their other genes might have been
rotten—they might have made them mean or selfish or stupid—but it didn't make any difference; their genes were
passed on to the next generation.

This is the ineluctable logic of evolution, the organizing
principle of life. Of all the things we know and don't know
about our ancestors, there is one thing we can be certain of,

going back to the beginning of humankind: they had sex and raised children. Otherwise we wouldn't be here. And when they joined together as man and woman they passed on the very genes that made them capable of and interested in having sex in the first place.

Evolution even helps explain some particularly enduring and annoying human traits, such as jealousy, possessiveness, and infidelity. When a man and woman mate, the genes have a vested interest in their staying together long enough to produce and raise a child. Paul and Madeleine's initial attraction hurled their bodies together, again and again, in order to reproduce. The relationship was just as passionately ruined because of infidelity. A man is furious when a woman strays because she could become pregnant by another male, which means the first male would spend all his energy raising someone else's genes. If a man strays, the woman is jealous because she needs help nurturing the child. On the other hand, a little infidelity isn't all bad, from the genes point of view, because it could help the man spread his seed and allows the woman to continue seeking the "perfect" mate with whom to reproduce. So a man wants to have many partners, while preventing those females from mating with others. A woman wants a faithful man, while reserving the right to seek sperm elsewhere.

Evolution explains *why* we are genetically programmed to feel so strongly about love, but it doesn't say anything about *how* the genes work. When a man and woman hungrily push their bodies together, they are not thinking about evolution, even if they're scientists. At the tremulous moment of orgasm, a man's thoughts usually are not about passing on his genes. If anything, in this day and age, both partners are probably much more concerned about *not* having a child than about reproducing. Such a sophisticated calculation could come only from the cerebral cortex, the most recent and highly evolved part of our brain. Since animals were feeling the urge to have sex long before they could think about such matters, even before they realized that sex was connected to reproduction, those "animalistic sex thoughts" must have been controlled by the limbic system, the more primitive part of the brain that serves as the seat of human emotion.

To guarantee their survival, the genes found a clever trick. Instead of appealing to our higher sense of calling, or our duty to continue the human race, the genes made sex feel good, real good. Genes code for millions of touch receptors in the genitals and for the nerves that connect them to the brain, the most important sex organ. In the somatosensory cortex, the part of the brain linked to the genital area is larger than any other, which is why the genitals are so delightfully, exquisitely sensitive to the touch. Other genes code for the flood of hormones that are released during pregnancy and at childbirth, infusing the mother with warm feelings toward her child. Still other genes, presumably in the primitive limbic part of the brain, help make us receptive to the social interactions and signs of mutual attraction that we feel instinctively and now call love.

THE ROOT OF OUR DIFFERENCES

As members of a species, men and women share the same goal of reproducing. As genders, however, men and women have very different ideas about how to achieve that goal. The difference is the difference between sperm and egg. Put simply, men behave like sperm, which are cheap and abundant; their best strategy when dealing with an egg is to find it, fertilize it, and forget it. Women behave like eggs, which are rare and valuable and which, once fertilized, require a substantial investment of time and resources in child care; their best strategy is to be picky, to find sperm from a man who will help with the child, and to ignore other potential mates.

From this basic biological difference springs all sorts of other differences between men and women, including the most basic traits we share with our animal ancestors, namely, who we are attracted to, how we attract them, and how we consummate our feelings. The biological difference between men and women also is at the center of more sophisticated distinctions, such as the difference between *Penthouse* and Harlequin romances, Mars versus Venus, combat boots versus pantyhose. There is no gene that prevents men from asking directions or requires them to hog the TV remote control; nor is there a gene that makes women want to talk about everything or to shop. People are individuals not totally defined by gender, and there is a large amount of free will involved in how men and women relate. That said, there are real, undeniable differences between men and women, and amazingly they come from a single genetic switch, in fact from a single gene.

The master gene is called TDF, which is named for the protein it codes for, the testis-determining factor. It is located, not surprisingly, on the Y chromosome, the only chromosome that men have but women don't. At the beginning of life for an embryo, there is no difference between a male and female, except for a mere streak that will become genitals. If the fetus is male, the TDF gene is turned on about eight weeks after conception, and differences begin to appear.

To carry out its male mission, the TDF gene activates another gene that makes Muellerian-inhibiting hormone, which keeps internal female organs from developing. The gene's second job is to activate a group of cells that synthesize testosterone, the male sex hormone that leads to the formation of the male genital tract. TDF works like a switch at a railroad yard: when it's absent, the train proceeds along the female track; when present, it switches development over to the male track.

TDF carries out its work in an instant. Once the train has been switched to the male track, the testis-determining factor is no longer important; in fact, by the time a child is born, nary a trace of the protein can be found. What give the "male train" its powerful momentum are hormones, specifically, testosterone and related male sex hormones synthesized by the testis. These hormones direct the development of the genitals and later cause facial hair and a deepening voice.

TDF is not the only ingredient necessary to produce a male. There are many "downstream genes" that also must switch on to keep development on track. In rare cases, the process doesn't work. The results can be confusing and at times heartbreaking.

María Patiño was a determined young woman and champion hurdler from Spain who planned on winning when she arrived in Kobe, Japan, for the 1985 World University Games. Her dream turned to a nightmare before she even made it onto the track. She flunked the sex test.

Sex testing for female athletes began in 1966 in response to rumors that some of the top Russian and Eastern European female athletes were really males. The original test was simple; the women paraded naked in front of a panel of gynecologists. If that had still been the test in Kobe, María would have passed because she appears to be a normal woman with breasts and vagina. But in order to avoid the embarrassment of the "nude parade," a new and supposedly more sophisticated procedure was introduced. A few cells were taken from the cheek, stained with a special dye, and examined under a microscope. If the cells had two X chromosomes, meaning a female, a dark spot would appear, but if the cells were XY, meaning a male, no dark spot would be visible. When María's test results came back, there was no dark spot. That meant she had an X and a Y chromosome and, to the judges, had to be a male and therefore disqualified from competing.

The news was shocking to María and her parents, who never suspected she was anything but a healthy young woman. Further tests revealed she had a rare condition, present in about 1 out of 20,000 (XY) births, called androgen insensitivity syndrome. There was a crippling mutation in the gene for the androgen receptor—the protein that senses the presence of the male sex hormones. María was born with a normal Y chromosome and TDF gene; her developmental train should have been directed in the male direction. But because there was no receptor to sense the androgen, it was a circular track. No external male features were formed, and the train was routed back to the female track. That's why her external genitalia and secondary sex characteristics were those of a woman. The only difference between María and any other woman was that María had internal testes and an incompletely developed vagina, which would seem irrelevant to her ability to leap over hurdles.

María's case is an example of how a little knowledge of genetics can be dangerous. The officials thought that XY must mean male, but in fact many other genes, most of them on other chromosomes, are necessary to distinguish a man from a woman. And since María was a woman under any definition except in the vocabulary of this particular gene, other genes must play a role in the behaviors that we consider to be female or male.

VENUS VERSUS MARS

Few dispute that men and women are different in their behavior as well as their anatomy. Even with clothes on, the distinctions are clear. Men are hard, aggressive, competitive, cold, promiscuous, lusty, bison hunting, car crazy. Women are soft, yielding, nurturing, emotional, monogamous, sensuous, berry gathering, lousy drivers, great talkers. In short, "women are from Venus, men are from Mars." The question is: are the emotional and personality differences between men and women hardwired in the genes or the product of culture and socialization? Are men nasty, horny, competitive brutes because they are taught to be that way or because they can't help it? Are women such fragile flowers because their mothers told them that's how to win a man's heart or because their brains were created that way? The debate over the origin of sex differences probably dates back to the first prehistoric man who tracked mud into the cave. Few questions have caused more controversy or produced more fuzzy and misguided thinking.

Interestingly, a single case from the medical files has been used to argue both sides of the issue. The story begins with the birth of a pair of beautiful, healthy twin boys. At eight months, one of the boys, known in the case history as

John, needed minor surgery to repair a fused foreskin. Incredibly, the surgeon botched the job and sliced off the tiny penis. The doctors convinced the parents that allowing the boy to grow up a eunuch would be too traumatic, so they proposed that the most charitable thing would be to finish the job. The parent's agreed, and the boy's testicles were removed. In further surgery, the doctors began the long process of constructing a vagina for the child, who was essentially born again as a girl. The parents began calling the baby Joan and bought her new pink clothes, frilly dresses, and dolls. A 1973 article by John Money, a pioneering sexologist, described how Joan was growing up to be a perfectly normal little girl, and the case seemed to prove that sexual identity could be determined by loving parents working with trained professionals. Children, it seemed, were born gender neutral and could be guided just as easily down a male or female path.

Recently two other researchers, Dr. Milton Diamond and H. Keith Sigmundson, decided to see whatever happened to Joan some 30 years later. Their discovery, published in 1997, turned upside down the theory of sexual identity. Joan, it turned out, did not become a happy little girl. She tore off her dresses and tried to urinate standing up. Her mother taught her how to use makeup, but she preferred imitating her father shaving his face. Other little girls also seemed to recognize instinctively that something was wrong; they called her "caveman" and refused to play with her. The parents, hiding the secret from everyone, saw no choice but to continue down the path they had chosen, and started Joan, at age twelve, on estrogen to give her breasts and other female characteristics.

By age 14, Joan was depressed. She knew something was

wrong, and she didn't want to go on living this way. Why wasn't she like other girls? Why did she want to play baseball with boys but couldn't imagine kissing them? Suicide was the only way out, she thought. Her parents knew the reason for her suffering and couldn't bear to keep it from her any longer. Her father tearfully recounted the story, even though he knew she might blame and hate them. But instead of being angry at her parents, Joan was relieved. "For the first time everything made sense, and I understood who and what I was," Joan told the researchers.

Joan then began the path back to her biological roots as John. The breasts were removed and a semblance of a penis was added surgically. Male hormones replaced female hormones. John embraced his new identity. "He got himself a van, with a bar in it," Diamond told the *New York Times*. "He wanted to lasso some ladies." At 25, he married a woman and adopted her children. Today, more than 30 years after a tragic and misguided detour, John says he has finally arrived at the place he was born to be.

The case of John is now used to show that sexual identity is at least partially complete at birth. John could only be happy as a man because he was born a man. No amount of socialization could change the way his brain had been formed in the womb. John knew instinctively that something was wrong with his gender, and it wasn't just that he didn't like playing with dolls. His whole identity was colored by gender, which means being male or female is part of the foundation of personality. The difficulty is in describing the differences between male and female. The anatomical clues are easy, but what does it really mean to be born a man or born a woman? How are men and women different and why?

WHAT MEN WANT IS NOT WHAT
WOMEN WANT

Although men and woman seem very different sometimes, they are, believe it or not, members of the same species and have the same long-term, evolutionary goal: children. But as in many things, they have very different ideas about how to achieve that goal. By looking at some of the different sexual behaviors and strategies of men and women, evolutionary psychologists believe they can point to the types of genes involved. For men, the optimal strategy to preserve the species is to have as much sex with as many different partners as possible. For women the optimal strategy is to be highly selective and to have sex only with a man who will commit time and resources to children. This is why young men obsess about having sex and young women worry more about relationships. Judging from the marketplace, men buy books and magazines about sex and how to find it; women are more interested in making romance "work," about "keeping" Mr. Right, and getting married. Interestingly, even homosexuals are not immune to this kind of sex difference. Here's a joke: What does a lesbian bring on the second date? A U-Haul. What does a gay man bring on the second date? What second date?

When a man cruises a bar, or a riverbank, or the plaza of a Mexican village, he is looking for youth and beauty, firm flesh, bright eyes, luscious lips, pert breasts. This might seem like something out of a *Playboy* fantasy, but Dr. David Buss and his colleagues have found it true in 37 different societies. In the United States, men seek women who are on average three years younger than themselves. As men age, the disparity grows: first wives are about three years younger, second wives are about five years younger, and third wives are about eight years younger than their husbands. In *The First Wives Club,* Bette Midler attends her son's bar mitzvah, to which her ex-husband brings his much younger girlfriend, and Midler asks, "What is she—a gift?"

Despite occasional exceptions like Cher, women prefer the opposite. In all of the 37 cultures surveyed, women expressed a desire for men older than themselves, ranging from two years older in French Canadians to five years older in Iranian women to several decades in the Tiwi tribe, a gerontocracy in which the elders control all wealth and prestige. The average difference worldwide was three years.

Here's another joke: Two 75-year-old guys are talking. One complains how he's not getting any. The other says he has a new girlfriend, 25 years old. "How'd you do that?" asks the first. "Easy," says the second. "I told her I was 95."

The joke reflects the theory that what women really want in a man is resources, the kind accumulated with age. The simplest resource is money. In our culture it is not uncommon for families to be pleased when a daughter marries a man from a "nice" family, meaning with money. But if a man goes after a woman with money he is a gigolo. Even if a woman has no conscious desire to have children, her genes are "telling" her to choose a man with the resources to raise them. When American women and men were asked in 1939 what they were looking for in a spouse, the women rated financial prospects twice as important as did men. Virtually the same result was found in 1956 and 1967, and again in the mid-1980s, after the supposed sexual revolution. What this means is that the best thing a man can pull out of his pants is a big, thick wallet. Are women looking for rich guys because that's what they learn from *Cosmo*? That seems unlikely since the same preference for men with money was echoed by women in 37 different cultures on six continents and five islands, in polygamous as well as monogamous societies, in the Third World as well as the richest countries.

The next best thing to money is power, which in our culture is judged by social and professional status. "Power is the most potent aphrodisiac," said Henry Kissinger, who ought to know; he had power and money and no shortage of women, despite looks that were not exactly made for the movies. Research backs up Kissinger's analysis. In surveys, women consistently rate status, prestige, rank, power, and station more highly than do men, and that's true across the world.

Men also seek resources, but a different type. What men really want is fertility. Since they can't judge fertility by looking, they use a surrogate marker: beauty. When men ogle a full pair of breasts, hard nipples pushing against a sheer blouse, they are not thinking, "My future infant son will grow strong sucking those." But their genes are shouting it to the heavens. Men love round and ample buttocks, not because they learned it from television, but probably because the shape is perfect for bearing children. Facial symmetry—which experiments have shown is a cross-culturally recognized sign of beauty—is a billboard that declares the person free of parasitic disease. Men's standards of facial beauty are remarkably similar even across racial and ethnic groups: the most attractive features are a small nose and large widely spaced eyes, just as much in Africans as in Japanese and in Swedes.

In America, there is no doubt about our obsession with beauty, and it's not new. In a 1950 survey of 5,000 college students about what they sought in a mate, the factor men endorsed by far the most frequently compared with women was physical attractiveness, and this has not varied during the past 40 years. The one physical thing women do appreciate is height, probably because it's a surrogate for physical prowess and thus ability to provide food and fight off enemies. This is why so many classified ads placed by women seek men over six feet tall, and why the ads placed by tall men do better than those by shorter men. The interesting question is what will happen to a species designed for physical tasks in the Stone Age that suddenly transforms its environment into an information-dominated world where physical prowess is less important. Will men with glasses—which could indicate extensive reading or computer work as a child and therefore intelligence—become more attractive?

Men's preference for beauty as compared with women's preference for status is illustrated by an experiment described by John Townsend of Syracuse University. More than 200 male and female college students were shown photographs of people and asked who they'd want to date. Some of the models were dressed in fast-food uniforms and others wore fancy professional clothing and Rolex watches. The men's willingness to date certain women and not others was based solely on physical attractiveness. If she was pretty they'd date her, even if she was working at McDonalds. But for the women, the suit made the man. Women were as willing to date the well-dressed homely guy as the fast-food beefcake.

Just as females of many species are attracted to a well-groomed male, they also are impressed if other females are interested. Looking at an example in animals, a male Trinidadian guppy in the company of a female is very attractive to other females, even if he wouldn't be by himself. For guppies, color is magic, the color of choice is orange, and the most orange male is most likely to be approached by a female. Dr. Lee Dugatkin of the University of Louisville positioned a tank of female guppies next to tanks of males. When the male tanks were stocked with females to give the impression that the males had companions, the females in the other tank preferred the males with girlfriends. Moreover, pigmentally challenged male guppies could overcome their genetic handicap by having a date. Even a pale male—24 percent less orange than average—improved his chances by being seen with a female guppy. The females weren't totally snowed, however, and if a male was 40 percent less orange, he was out of luck no matter how many females were swimming in his wake.

This same strategy works, at least to a certain extent, in humans. Every guy knows that the best way to attract attention is to walk into a party with a gorgeous woman, as long as they're not married. That's because commitment is one of the most important qualities that women seek in both casual and long-term sexual partners. From the point of view of a woman's genes, sex is sex, whether it's motivated by a flighty passion for a handsome stranger or a sense of marital obligation with a husband she keeps around like an old shoe. If a baby can be the end result, then the partner's level of commitment matters greatly. By contrast, men value commitment and sexual fidelity in long-term partners, but find those values a turnoff for more casual sex.

Could something as culturally contextual as "commitment" really be biological? Although there's no proof in humans, an animal called the vole provides a fascinating example of how genetic makeup and brain chemistry can control behavior. Voles are little rodents similar to field mice. There are two closely related species, prairie voles and montane voles, that couldn't be more different when it comes to commitment. Male prairie voles mate for life and will fight off intruding males tooth and paw. By contrast, male montane voles mate with everybody. Why would two types of voles, so similar in most respects, be so different when it comes to commitment?

Researchers Thomas Insel and C. Sue Carter at the National Institute of Mental Health's Lab of Neurophysiology found the two species had remarkable differences in the pattern of receptors for a peptide hormone called vasopressin. When vasopressin was blocked in the brains of normally faithful male prairie voles, they mated around the colony and failed to defend their females from other males. It was as if the most faithful husband had suddenly plunged into promiscuity. The chemical reaction worked the other way as well. When a male prairie vole was living with a female, its brain produced more vasopressin than when it was alone. No change was seen in the male montane voles, nor in females from either species. Thus, a small genetic difference between the two species resulted in a large difference in mating behavior, simply by changing the amount and distribution of a brain hormone.

Back to our field research in the singles bar, where the guys crave luscious young babes and the women are holding out for tall rich gentlemen. Now it's closing time, people are lined up for their coats, and you are still alone. What will you settle for? For men the answer is simple: just about anything still walking. In one survey, 75 percent of male college students said they'd be glad to have sex with an attractive woman even if they knew *nothing* about her. But when women were asked if they would have sex with a handsome stranger, 0 percent said they'd accept. When asked what mattered in a sexual partner, the men were less demanding than the women in just about every category, including education, intelligence, charm, sense of humor, and personality. Nor did the men care about ignorance, low social status, drinking too much, selfishness, or even mental instability. While the women wanted to know at least something about the men before they agreed to sex, the men seemed to be concerned only about a functioning vagina.

All of these differences in what men and women want—beauty versus money, youth versus age, availability versus commitment—seem like they could be genetic. They are observed across human societies, so they can't be purely cultural. In many cases they are seen in other species, so they are probably evolutionarily conserved. And they are different from one person to the next—not all men are pigs, not all women prigs—hinting at individual genetic differences. But seeming is not proving. We have no idea what these genes are, how many there are, what they do, or how they are differently expressed in males and females—all of which is necessary to prove a genetic basis. Still, there are two areas of sexual behavior where specific genes have been identified. Not surprisingly they have nothing to do with genitals and everything to do with the brain.

GETTING YOUR KICKS FROM SEX

President Calvin Coolidge is not usually thought of as a major contributor to sexuality research. Nevertheless his name is associated with an important phenomenon. The story is that Coolidge and his wife were inspecting a government farm. While the president was off somewhere, Mrs. Coolidge observed a rooster mating with a hen in the chicken coop. "How often does he do that?" she asked. "Dozens of times a day," said the guide. "Please mention this fact to the president," said Mrs. Coolidge. Later, when the president passed the chicken coop, he was told of the rooster's prowess. "Always with the same hen?" the president asked. "Oh no, a different one each time," said the guide. The president said, "Please tell *that* to Mrs. Coolidge."

The Coolidge effect is the interest of males in a variety of sexual partners, and it is well documented in humans. According to *Sex in America: A Definitive Survey*, a large and careful study from the University of Chicago, the typical American man has six sexual partners over a lifetime, while a woman has two. Those numbers are mathematically impossible, unless there is a hidden minority of women who sleep with many men, but they certainly reflect what men would like to be true.

Our culture reinforces the idea that a man will be attracted to many women. This is why bikinis are used to sell "male" products such as tires, and not blenders. But don't think this pattern is merely an artificial creation of Madison Avenue; it's more like a recognition of something found across human cultures and in other mammals. David Buss, in his book *The Evolution of Desire* mentions the Mehinaku men of the Amazon, who call sex with a spouse *mana* or flavorless, but sex with a lover *wirintyapa* or delicious. He also quotes an Indian man who says, "You don't want to eat the same vegetable every day."

Nor is the Coolidge effect limited to our species. A ram, for example, will mate as many times as it finds new females. But once he's had a particular ewe, she becomes less interesting. This is not because he knows the ewe has been inseminated; he doesn't care how many other rams she's had, as long as she's new to him. Even if a ewe he's had is covered up with canvas and "disguised," or taken out of the pen and returned so he thinks she's new, the male is not fooled. What the ram wants is novelty.

Male interest in a variety of sexual partners shows three of the hallmarks of genetic traits: it's true across cultures and time, it's found in other species, and it shows individual variability. But what is the gene?

The clue came from studying personality, specifically the trait called novelty seeking, which means finding pleasure in new, varied, and intense experiences.

Several studies have shown that high novelty seekers satisfy their need for change and variety by having a large number of sexual partners. For example, Anthony Bogaert and William Fisher at the University of Western Ontario found that a novelty-seeking score was a better predictor of the number of sexual partners than was any other factor, including physical attractiveness, masculinity, age, or general interest in sex. The more a person was a thrill seeker, the more partners he had.

In another study, Marvin Zuckerman and colleagues studied college students during the so-called sexual revolution of the early 1970s. High novelty seekers said it was fine to have intercourse with someone they just met, even if they weren't sure they liked each other. Low thrill seekers were more likely to endorse sex only when deeply in love, and preferably married. Thrill seekers viewed sex as a "game," a pleasurable form of recreation, while low scorers saw sex as an expression of emotional commitment.

Novelty seeking also affects the kind of sex. Zuckerman's college survey showed that men who scored high for thrill seeking had a greater extent and variety of sexual activities than low thrill seekers. High thrill seekers were much more likely to engage in oral sex and positions other than missionary style. The link between thrill seeking and sexual voracity is not about sex drive, however. Zuckerman's study showed that high thrill seekers didn't masturbate any more than low thrill seekers, but they did have more partners. Thrill seeking is about how and whom we have sex with, not how often.

Since all these studies pointed to a relationship between novelty seeking and sexual behavior, and since we had previously shown that novelty seeking is in part mediated by the D4 dopamine receptor, we were poised to ask an interesting question: does the D4 gene influence the number of sexual partners? Because we already had the data, it took only a little number crunching to get an answer.

Sure enough, there was a correlation between the D4DR gene and number of sexual partners—but in an unexpected way. Looking first at the straight men, we found that the ones with the long form of the D4DR gene, the high novelty seekers, had slightly more female partners than those with the short form, the low novelty seekers. The difference was small, though, and not statistically significant. But the study got more interesting when we asked how many other men they had slept with. Although these were straight men, some had slept with another man, usually just once and when they were young. But here was a very strong correlation to the D4DR gene. Straight men with the long gene, the high novelty seekers, were six times more likely to have slept with another man than those with a short gene. About half of the long gene subjects had ever had a male sexual partner compared with only 8 percent of the short gene men.

Just the converse was true for gay men. As expected, the gay men had more male partners than the straight men did female partners, probably because in the gay world the Coolidge effect is universal, and the D4DR gene had the expected effect. However, the effect of the gene was much stronger for the number of female partners of the gay men. Those with the long, high novelty-seeking form of the gene had sex with more than five times as many women as did those with the short, low novelty-seeking form. Although gay men may have had sex with women in part because of social pressure, it seemed that a desire for new experiences also played a role.

These results show that the D4 dopamine receptor gene does influence male sexual behavior, but indirectly. For a straight man, sleeping with another man is about as novel as you get. For a gay man, having sex with a woman is equally unique. Does this mean that D4DR is a "promiscuity gene" and that an errant husband can tell his wife, "I couldn't help it, it was genetic?" Of course not. A gene doesn't make a person commit adultery. It simply determines the way certain brain cells respond to dopamine, which in turn influences a person's reaction to novel stimuli. How a person reacts to that stimulus is more a matter of character than of temperament.

How Much Sex Is Enough?

In the movie *Annie Hall,* Woody Allen tells his psychiatrist that he and Annie have sex "hardly ever, maybe three times a week." Annie tells her psychiatrist that they have sex "constantly, I'd say three times a week." According to *Sex in America,* 54 percent of men think about sex at least once every day, but 67 percent of women say they think about sex only a few times a week or month. An even bigger difference is found among individual men and women, however, so there is actually more overlap than difference between the genders. For example, some people—men and women—never have sex during their whole lives, while for others once a day is barely enough.

How often a person has sex depends on many variables and changes frequently. People have more sex when they are younger, and the frequency depends greatly on the availability of a partner. Despite the popular notion of "swinging singles," married people have the most sex. About 40 percent of married people have sex twice a week compared with 25 percent of singles. For married couples, probably including Mr. and Mrs. Coolidge, frequency tends to decrease over time. On the other hand, there are bursts of sexual activity that can occur at any stage of a relationship, such as during vacations. Even taking into account all the outside influences on sexual frequency, it's clear that some people just gotta have it more often. And despite changes that will occur as a person ages, marries, divorces, or has periods of stress, the level of "sex drive" is a relatively stable part of a person's makeup.

What genetic brain mechanism could account for such stable differences in libido? This time the clue came from pharmacology, specifically from studies of the antidepressant drug Prozac. One of the few negative side effects of Prozac is the loss of sexual desire. In a way, this shouldn't be surprising since sex is often used to relieve anxiety or to blow off steam. In men, Prozac's effects are not just psychological; men may have trouble getting erections or ejaculating.

We knew that Prozac works on the brain transporter

serotonin, and we had discovered that some people have a version of the serotonin transporter gene that works like natural Prozac to reduce anxiety and depression. The obvious question was: does the genetic Prozac also reduce sexual desire? Was this part of the body's sexual control that determined how often people jumped into bed?

Again, it was fairly easy to use our male research subjects to compare the frequency of sex with the type of gene. There was a significant correlation: men with the high-anxiety form of the serotonin transporter gene had sex more often than those with the low-anxiety form. The most sexually active men, those who had sex at least once a week, were 1.9 times more likely to have the high-anxiety genotype than those in the less-frequent-sex group. This was true regardless of age, education level, ethnic group, or sexual orientation. There were many exceptions because the gene accounts for only a small portion of the variability in sexual frequency, just as it does for anxiety and depression. This is not the master switch on the sexual thermostat, but it does appear to be part of the machinery. It also explains why a gene that causes anxiety, pessimism, and the blues is present at high levels in the human population. If the very common high-anxiety form of the gene "causes" people to have more sex, it will be passed on to future generations regardless of how they feel about it.

SEXUAL ORIENTATION

Probably the best known, and certainly the most controversial, work on the genetics of sex is the discovery of the so-called gay gene by my lab in 1993. Contrary to the focus of the news reports, the discovery is important not so much because of what it says about homosexuality in particular, but rather because of what it says about sexual orientation and sex differences in general. Sexual orientation is, after all, the most fundamental of the differences between the sexes. In all cultures and at all times in history, nearly all men have been attracted to women and nearly all women have been attracted to men. Because genetics is by definition the study of inherited differences, the best way to understand the genetic basis of sexual orientation is to compare people who vary in their attraction, which means studying gays.

The tricky part is defining and measuring sexual orientation. Some scientists want to measure only actual behavior: does the person have sex with men or women or both? Others, believing that society prevents us from acting on our "real" desires, would add fantasy to the equation: does the person feel attracted to men or women? Another way to measure sexual orientation is simply to ask: are you gay, straight, or in-between?

The traditional method of measuring orientation is the "Kinsey scale," named after the pioneering sex researcher Alfred Kinsey. The Kinsey scale is a yardstick from 0, for exclusive heterosexuality, to 6 for exclusive homosexuality. The numbers in between represent various degrees of bisexual behavior and fantasy. Probably the best way to measure sexual orientation—which is the way we used in our studies at the National Institutes of Health—is to measure at least four different things: attraction, fantasy, behavior, and self-identification. A person's total score is the average of the four areas. So a man who has slept only with women but admits to occasional fantasies about men would score between 0 and 1 overall.

Most people, men and women, will score close to 0 because nearly all people are heterosexual. Some people—estimates vary but not more than a few percent—will score higher because they are bisexual or homosexual. The interesting thing, however, is the subtle but important difference in the distributions of the scores for men and women. This difference is key to understanding the underlying basis of human sexual differences.

The difference is that most men are at one end or the other of the Kinsey scale—either they're straight or they're gay—while women are much more spread out along the scale—many of them show some degree of bisexuality, either in their behavior or, more frequently, in their thoughts. An example is a study of 4,903 twins in Australia in which the majority of both the men and the women identified themselves as predominantly or exclusively heterosexual (Kinsey 0 or 1). However, for the men there was a "valley" in the distribution in the bisexual score range of 2 to 4, then a "peak" in the predominantly or exclusively homosexual range of 5 to 6. But for the women there was a gradual tapering off of the distribution in the higher Kinsey score range. There wasn't any peak at all in the predominantly or exclusively lesbian range; most of the nonheterosexual women were in the intermediate, bisexual range.

The results showed that sexual orientation in men was similar to being right-handed or left-handed: there wasn't much room in between. Most of the men were heterosexual, while a minority were gay. But for women sexual orientation was more like height, with a wide range of options. This isn't to say that there aren't *any* bisexual men or *any* exclusive lesbians; just that there are *relatively* more bisexual women than men, and *relatively* more exclusively gay men than women.

A second difference is in consistency of orientation, meaning that individuals have the same score for attraction as they do for behavior and the other categories. This came through clearly during the hundreds of interviews we conducted for the gay gene study.

Christine was an intelligent, articulate young woman, a professional with a senior position in a large corporation. She had no problem talking about her sexual history, likes, and dislikes until one question caused her to become tense and agitated: "When you imagine having sex with someone else, for example during masturbation, who do you fantasize about?"

Christine blushed, looked down at her feet, then softly replied: "This is all confidential, right?" She was assured that no one would ever learn her identity. "Well," she continued, "It's usually about men. That's the only way I can get an orgasm."

What was surprising about Christine's reply was that she described herself as a lesbian since the age of 17. She had been with the same female partner for eight years. All of her sexual experiences, besides a few dates in high school, had been with other women. When asked about sexual attraction—who she might actually want to go to bed with, rather than just think about—she said only women. On every Kinsey scale measure except fantasy, she was clearly a "6."

Christine is not unique. Many women show a similar fluidity between fantasy, attraction, behavior, and how they define themselves. Nor was her reply proof of the old myth that what lesbians really want is a man, since straight women showed the same fluidity in their responses. For example, one happily married heterosexual woman who had never had any sexual experience with women reported that her favorite fantasy was "being under a pile of women who beat my face with their breasts."

Men are on average more consistent, even rigid, in their sexuality. Men who identify as heterosexual virtually always are attracted to and fantasize only about women, while those who call themselves gay think and dream almost exclusively about men. Of course when it comes to actual behavior there are exceptions. For example, many gay men have had sexual experiences with women, but usually because they felt they were expected to by their families and society. Many heterosexual men have at least some sexual experience with other men, but most often it is a brief and youthful experimentation or because they are in a situation without women, such as in prison.

A third difference between men and women is what happens over time. Men on average stay pretty much the same, whether straight or gay, during their entire lives. Although men usually don't acknowledge to others, or even to themselves, that they have a homosexual orientation until late adolescence or early adulthood, once that has occurred they are unlikely to change. Moreover, both gay and straight men can usually trace back their attractions to early childhood, even as early as four or five years of age. Early crushes or puppy love for gay boys are often with other boys or men.

Women are different. Margaret, in her late sixties, said she had been married twice and had always enjoyed sex with her hubands. She had never had any interest in sleeping with a woman, nor any other type of experience. She was a perfect "Kinsey 0," an open-and-shut case. But as she was preparing to leave the office, Margaret said, "There's one thing you didn't ask me, Doctor."

"What's that?"

"You didn't ask me about the future. You know I may be 68, but I'm still interested in sex. And you may not know this, but the men in my age range are *pathetic.* So I assume that my next lover will be a woman."

And then she stood up and left, leaving behind one rather astonished researcher wondering how to incorporate that brief but revealing conversation into the neat and tidy columns of Kinsey scores.

As the research progressed, it turned out that Margaret was not unusual. With researcher Angela Pattatucci, we interviewed other women who had described themselves as lesbians but then married men and women who "became" lesbians later in life. Still others had gone back and forth two, three, or more times during their lives. For some women, sexual orientation seemed as fluid as their weight—and they were yo-yo dieting.

The research showed that male sexual orientation had many of the characteristics of a genetically influenced trait: it was consistent, stable, and dichotomous, meaning men were either gay or straight. By contrast, female sexual orientation looked more soft and fuzzy, less hardwired: it was variable, changeable, and continuous, meaning lots of women were somewhere between gay and straight. Just because a trait looks genetic, however, doesn't mean it is. We needed to look at twins, families, and DNA.

During the past 40 years, more than a dozen twin studies of male sexual orientation have been described, and the pattern is the same. The genetically identical twin of a gay man has a greatly increased chance—though not a 100 percent chance—of also being gay, which is higher than the rate for fraternal twins, which is still higher than the rate for unrelated people. This is just the pattern expected for a trait that is influenced—but not strictly determined by—genes. Averaging all the studies to date, the heritability of male sexual orientation is about 50 percent. That means that being gay is about 50 percent genetic and 50 percent from other influences, a ratio found in many other behavioral traits.

So what about the missing 50 percent? Why can one man be gay even if his identical twin is not? The answer is not yet clear, but it could be biological, such as different hormonal exposure in the womb, or because of unique life experiences. One thing that is *not* terribly important is how the boys are raised, specifically the shared environment provided by parents. In the most careful twin study to date, the best mathematical estimate for the shared environmental component of variance was 0 percent. In other words, if two boys are raised in the same home and one turns out gay and the other straight, the difference was not caused by the general parenting style.

For women, the degree of genetic influence is more mysterious, partly because there have been fewer studies but also because sexual orientation is more fluid. The best recent study suggests that female sexual identification is more a matter of environment than of heredity. The new study was conducted in Australia by Nicholas Martin, a highly respected behavioral geneticist, and Northwestern University's Michael Bailey, who has done most of the recent population genetic work on sexual orientation in the United States. Instead of recruiting subjects through advertisements, which leaves the door open to bias, they used a national registry sample of 1,912 female twins from 17 to 50 years of age. The rate of lesbianism was higher in the twins of lesbians than in the twins of heterosexual women, but there was no difference between identical twins and fraternal twins, meaning genes were not a factor. The results showed that for women the main influence on sexual orientation was the shared environment—being raised in the same household by the same parent—while genes seemed to count hardly at all.

GAYS IN THE FAMILY

Genetic traits such as eye color run in families, but so do environmental traits, such as language. The secret of genetic research is how traits, including sexual orientation, move through generations, whether they are in the blood or in the atmosphere.

Earlier studies, especially the work of Professor Richard Pillard, a Boston psychiatrist, had found that the brother of a gay man had a fourfold increased chance of also being gay. But Pillard's study could not say *why* the brothers were gay. Since they were raised together, it could have been either genes or the family background. One clue would be other relatives, such as uncles and cousins, who would share genes but not a roof. That's what we set out to examine in 1992 as part of a study initially designed to look at the role of genes in Kaposi's sarcoma, a virulent type of cancer that was striking gay men with AIDS.

We started by interviewing gay men about sexual orientation in their families. Then, whenever we could, we interviewed the family members as well. When we analyzed the first series of results, a striking pattern emerged. Not only did gay men have more gay brothers—the same result found by Pillard—they also had more gay cousins and uncles. Since these relatives were raised in different households, indeed mostly in completely different parts of the country, the evidence for genes increased.

Even more striking was that the gay uncles and cousins were concentrated on the mother's side. For a gay man, his mother's brother had an increased chance of being gay, but his father's brother did not. For cousins, only the son of the mother's sister was especially likely to be gay; the son of the mother's brother, or of any cousins on the father's side, had just the usual chance. That was such a striking finding that we decided to check it using a different group of subjects, in this case the families of gay men who had been specifically recruited because they had multiple gay relatives, usually brothers. Once again, the pattern held up: most of the gays were through the mother's relatives.

We were excited that we were on to something, but we knew there are traits handed down by mothers that have nothing to do with genes. For example, being Jewish passes through the mother's side of the family because that's Jewish law. So do chocolate cookie recipes, for purely cultural reasons. But to a geneticist, all these gay men on the maternal branches of family trees meant one thing: a gene on the X chromosome. Because males always receive their X chromosome from their mother, any gene on the X chromosome is passed through the mother. Indeed, the pattern of male homosexuality looked similar to the patterns for color blindness or hemophilia, two classic X-linked traits. The inheritance was less strong than for a purely genetic trait, but that wasn't surprising because sexuality is so complex that there were likely to be multiple genes and strong environmental factors. Sexual orientation compared with something purely

genetic such as color blindness was like a water color reproduction of an oil painting: the pattern was the same even if
the colors were diluted.

Meanwhile, Angela Pattatucci was building family trees
for the female subjects, and just like the gay men, the lesbians
clustered in families. For example, the sister of a lesbian had a
6 percent chance of being a lesbian, which was about six
times higher than the baseline rate. But the surprising thing
was that the highest correlation was for lesbian mothers and
daughters. The rate was a whopping 33 percent, meaning that
the daughter of a lesbian had a one-in-three chance of also
being a lesbian.

Genetically speaking, this result was impossible. This
was the one family pattern that could *not* come from genes.
There was no genetic model that could explain how a parent
and child could be more similar genetically than two sisters.
The opposite pattern—when sisters were more similar than
mothers and daughters—was easily explained by recessive
genes. But the pattern we observed could mean only one
thing: being a lesbian, or a nonheterosexual woman, was
"culturally transmitted" not inherited.

Exactly *what* was being "transmitted" we couldn't figure out. The most obvious thought was that mothers were
somehow "teaching" their daughters to be lesbians. Maybe if
a young girl had a lesbian mom, she would imitate her as she
was growing up. But in many of these families, the daughters
actually came out *before* their mothers did. For example, one
mother described how her daughter came out to her by
bringing home a girlfriend from college during Christmas
break. Two years later, the mother had divorced her husband
and replaced him with a girlfriend. That Christmas holiday
all four women celebrated at a gay bar.

Perhaps what was being transmitted wasn't attraction to women, but rather a certain way of dealing with life. Perhaps it was a willingness to listen to one's own heart rather than the dictates of society, or an openness to new feelings and new experiences. It could even be an attraction to people based on who they are as individuals rather than on the shape of their genitals.

If these results hold up under further testing, it would appear that whatever is being transmitted to lesbians is fundamentally different from what is transmitted to gay men. It's more environmental than genetic, more nurture than nature. Why should there be such a deep difference?

Some will argue that any difference between male and female sexual orientation is all cultural rather than biological. Women are more likely than men to be bisexual, to change their sexual orientation, and to be influenced by their mothers, not because of genes but because of society. Women are encouraged to be in touch with their feelings, men are supposed to suppress any gay thoughts. Boys are ridiculed for being effeminate, but tomboys are spunky. These things are true of our culture, but why? Maybe it's because that's the way men and women really are, and we are just setting up mores to support the reality. In this way culture would support the biology, and the nature holds up the nurture.

LOOKING FOR GAY GENES

Once it seemed clear that male sexual orientation had a genetic component, the next—and still ongoing—step was to figure out what the genes actually were, to map them to the chromosomes, and ultimately to isolate them and see what they coded for or made in the body. Because we'd found a pattern of maternal transmission, we thought the X chromosome would be a good place to start searching.

With thousands of possible genes, we focused first on big chunks of DNA rather than individual genes. Like throwing darts in the dark, it was easier to hit the wall than the dart board. We knew it would be a difficult search since sexual orientation, even in men, was only partially genetic, and since the genetic component probably involved many different genes, not a single switch that said gay or straight. What we were looking for was one of many different factors that influenced sexual orientation, not a single, all-powerful "gay gene." Despite the hype, there is no such thing.

In linkage analysis, we compare DNA to find similarities in families. For example, if two people in a family are color-blind, they probably will share that piece of DNA that controls color vision. We weren't looking for specific genes but markers or signposts that showed whether two gay brothers inherited the same or different regions of the X chromosome from their mothers. If there really was a sexual orientation–related locus, then gay brothers would tend to inherit the same markers in the vicinity of this gene. In other words, if they had the same dart board, they'd also have the same piece of wall containing it. By random chance, 50 percent of them would have the same marker anyway, but if more than 50 percent shared the marker it would be significant.

To increase the sensitivity of the search, we used two research tricks. The first was to concentrate, at least at first, only on brothers who were definitely gay. The reason was that if someone said, "I'm gay," they were almost certainly telling the truth. There was no advantage to declaring homosexuality if you weren't gay. In contrast, if someone said, "I'm straight," they probably were telling the truth, but not always. It was possible they didn't want anyone to know they were gay, or they didn't want the information on a government record, or they hadn't accepted it themselves.

The second trick was to concentrate on families with the right kind of inheritance. We wanted to find something on the X chromosome, so by definition it had to come from mother to son. If either the son or the father of one of the gay brothers was also gay, they were excluded. There was no way a father could pass his X chromosome to his son, so if they both were gay, it must have been for another reason. Families with multiple lesbians also were excluded, since the pedigree pattern applied only to males, as were families with more than two gay brothers because they were so unusual. In short, we narrowed our search to have the best chance of finding something.

Our critics pounced on this research design and called it biased. They were absolutely correct. Our aim was to see if there was *anything* on the X chromosome that was related to being gay, not to measure the effect of that gene in the population at large. If there were no such gene, it wouldn't make any difference how we selected the families—we could look only at gay brothers with the middle name "Snitzleberg"— because that wouldn't affect their gene sharing on the X chromosome. In the same way, a naturalist searching for a rare tropical butterfly shouldn't waste time looking in Manhattan; but even if he goes to the perfect rain forest at the perfect time with the best equipment, he can't possibly find something that's not there.

As it turned out, we got lucky. Looking at 40 pairs of gay brothers with 22 different markers, we found linkage in a region called Xq28, located at the very tip of the long arm of the X chromosome. In that region, 33 out of the 40 pairs were concordant, or the same, for a series of five closely spaced markers. That showed 83 percent sharing, which was significantly higher than the 50 percent level that would have been expected if there were no connection to sexual orientation. When a statistical analysis showed it was unlikely that the results were a fluke, we decided to publish a paper about the finding, as much to stimulate further research as to make any sort of definitive claim.

The paper, with the catchy title "A Linkage Between DNA Markers on the X Chromosome and Male Sexual Orientation," was published in *Science* in July 1993. Rarely before have so many reacted so loudly to so little. The phone rang off the hook with calls from reporters; there were TV cameramen lined up outside the lab; the mailbox and e-mail overflowed; and T-shirts appeared that declared, "Thanks Mom, Love the Genes." Some conservatives called the study a government conspiracy to promote homosexuality, others were delighted at the mistaken idea that we had developed a gene test to ferret out homosexuals. One tabloid bannered a "cure" for homosexuality in a nasal spray. There were gay activists who rejoiced that homosexuality was "natural," others who accused us of being fascists intent on a gay holocaust. There were scientists who praised the rigor of the methodology and modesty of the claims, others who tried to discredit it. This was quite a strong reaction to modest scientific results: a finding that 33 instead of 20 pairs of middle-aged, white, gay brothers had the same tiny bits of junk DNA on one part of their X chromosome.

Among all the opposition, most of it because of the imagined implications of the results rather than the results themselves, two criticisms had merit. One was that the results had been found in only one experiment on one particular population of gay men. Given the poor track record of some early behavioral genetic studies, we wanted to repeat the experiment to be sure we were right. Chavis Patterson, a psychology student, collected a new group of families with two gay brothers who fit our selection criteria. Then we tested their DNA to see if there was any excess marker sharing in the Xq28 region. Indeed there was. This time 22 of 32 pairs, or 67 percent, shared markers. We gave the data to two expert statisticians, David Fulker and Stacey Cherny at the Institute of Behavioral Genetics at the University of Colorado in Boulder. We gave the same data to Leonid Kruglyak at the Whitehead Genome Center at MIT in Boston, who had developed a different analytic technique. In the end, they came to the same conclusion: there was a significant linkage to Xq28, and the results of the two experiments were statistically indistinguishable.

The second criticism was that we looked only at gay brothers, not at heterosexuals. In other words, we didn't check to see if the straight brothers lacked the "gay gene." There were solid statistical and epidemiological reasons for doing the initial experiment on gay brothers exclusively, and if I had to do it again I would use the same strategy, particularly since none of our critics has come up with a better method. Nevertheless, we were curious, so in the second study we also included the heterosexual brothers of gay men. We collected seven new families, plus four from the earlier study. As we suspected, most of the straight men had different markers from their gay brothers. Our statistical experts estimated that the degree of DNA sharing of the straight brothers with their gay brothers was 22 percent, significantly lower than the 50 percent expected by chance. This was another independent confirmation that Xq28 was involved in sexual orientation—straight as well as gay.

The evidence is compelling that there is *some* gene or genes at Xq28 related to male sexual orientation. What still is needed is corroboration from other laboratories. That hasn't happened yet, and the progess is depressingly slow. Four years after our results were published, no other group has published a single scientific manuscript on this topic. If we had discovered a gene for something less controversial, say diabetes or schizophrenia, scientists would have been fighting to replicate it, or shoot it down.

One study did claim a failure to replicate our results, but the research was so different it's hard to tell if it offers evidence for or against our findings, or any evidence at all. George Eberes and George Rice in London, Ontario, Canada, presented their preliminary findings at the Academy of Sex Research in Provincetown, Massachusetts, in 1995. They interviewed 182 families with two or more gay brothers and found significantly more gay uncles on the mother's side than the father's side, which would support our finding. However, when they looked at DNA—the critical part of the experiment—only 41 of these families (less than a quarter of the sample) were actually analyzed, and these happened to be mostly with paternal gay uncles. By definition, these families could not show X linkage because of the paternal connection, and they did not. That's not a refutation of our results; it's actually a back-handed confirmation, albeit not a significant one.

The final proof will be discovering the gene. Our experiments narrowed the search to an area with several hundred genes, equivalent to limiting a search for a needle in a haystack in Kansas. We do *not* expect to find a gene that is the same in every gay man—we already know that sexual orientation is more complex than that—just one that is correlated to sexual orientation. How might such a gene work? Perhaps it makes an enzyme that controls sex hormone metabolism in the developing brain. Perhaps it makes a growth factor that builds specific brain circuits. It could play a role we've never dreamed of since we know so little.

Some argue that the whole idea of a "gay gene" must be wrong because homosexuality is "unnatural." They believe homosexual behavior goes against evolution because it contradicts the real purpose of sex, which is reproduction. Or, in the words of a fundamentalist Christian critic who doesn't believe in evolution, our research must be wrong because it goes against the basic principles of "animal husbandry."

The critics raise a good question: how can a gene that leads to sexual behavior that is nonreproductive survive the rough-and-tumble of evolution? Why hasn't it been bred out of the human race? One reason is that homosexuality doesn't make reproduction impossible, and gays do have children. A second reason is that some heterosexual men have the presumptive "gay gene," which can be passed along to their children. On the other hand, even if the gene caused only a slight decrease in average reproductivity, it would die out unless something else were keeping it in the population.

This paradox has led to many theories of how a "gay gene" might actually be adaptive. One theory, although not a good one, is that it might be useful to the species because it prevents overpopulation. This is a poor theory because genes act at the level of individuals not groups. Others have suggested the gene might be passed along indirectly because homosexuals help their heterosexual relatives to raise children.

The simplest explanation comes directly from one of the most interesting results of the research itself: the gene only works in men, not women. We wondered whether the gene might have a different role in women, so we compared the mothers and sisters of our research subjects who were either linked or unlinked for Xq28. There was no difference in the number of their children or how often they had sex, but the women with the gay version of Xq28 did have one intriguing difference: they had begun puberty an average of six months earlier than the other mothers. Although the result is highly preliminary, it will be interesting to see if the gene somehow lengthens the reproductive span in woman, allowing them time to have more children.

GENES OF LOVE

Everyone, gay or staight, feels the tug of genes involved with sex and love, from the sharp pangs of puberty, to the defining role of gender, and the fierce, protective feelings of a parent for a child. Genes for gender and sexuality influence who we are and who we love.

In the case of Paul and Madeleine, the young lovers who began the chapter, mere DNA could not have predicted their intense romance and equally intense separation. We say people are "made for each other," and perhaps there is some human ability that helps us find a good mate. But the chance meeting of Paul and Madeleine was just that, chance. So many tiny pieces of the puzzle had to fall into place for them to be in the same bar at the same time and to meet. A single flat tire, a watch that ran too slow, a storm—an infinitely large number of factors—could have changed that day enough that the meeting never would have occurred.

On the other hand, genes made Paul a man and Madeleine a woman, thus making possible their sexual attraction. They met at the age of maximum reproductive potential, a time when biology throws young people at each other with enough force to tilt the Earth. Both had genes that made them attracted to the opposite sex; perhaps a simple variation could have made Paul sexually attracted to Madeleine's brother. Genes also might have influenced how the relationship developed. Was Madeleine a genetically driven thrill seeker? Did she pursue the ship's captain because she craved a new high? Or was it harm avoidance that caused Paul to flee at the first sign of trouble instead of sticking with Madeleine when she was confused? We are products of our genes, so it is natural that our relationships, too, will fall under the sway of DNA.

SIX

THINKING

Inheriting Intelligence

I think, therefore I am.
—RENÉ DESCARTES

I think I can, I think I can.
THE LITTLE ENGINE THAT COULD

Five-year-old Nicky lived in a New York City tenement with his two older brothers, an unemployed father, and a mother with a white-hot temper. His father never learned to speak English; and when a modest fruit stand failed to earn enough to support the family, the father was forced to look for unskilled day jobs whenever he could find them. At night the father sometimes drank too much and fought with Nicky's mother, their angry shouts flying out the open windows onto the busy street below. Once, when things were especially hard financially, Nicky's mother forged a check to buy groceries and was arrested. As a consequence, social workers came to the apartment, accused the parents of neglect, and took away the children.

The boys spent the next half-dozen years institutionalized in foster care. When they were allowed to return home, they found their mother living with three other sons from a previous marriage she had failed to mention. With six boys in the house, Nicky's father dropped by only occasionally and usually left after fighting with Nicky's mother. Margaret, the mother, was concerned mostly about paying the rent and buying enough food for everyone, so the boys dropped out of school and went to work to help pay the bills. All except Nicky, who was allowed to stay in school.[1]

Nick, a grown man now, admits that he was scarred by his rough childhood and unstable situation at home. "You never really get over foster care," he says. Survival itself was a challenge for a boy in the ghetto, where unemployment, violence, and crime were defining parts of the environment. Onto this hard soil, two struggling parents dropped the babies they produced. The mother and father probably didn't know much about theories of child development, they hadn't read all the parenting books, and they didn't enroll their boys in the best schools. Nick and his brothers were exposed to the harshest possible environment.

Somehow, Nick survived. Not only did he survive, he continued his education and excelled. He went to law school and became a respected lawyer making $400,000 a year. He became a man known for being intelligent, well spoken, and politically astute. But in one sense, he never did recover from his childhood and now in his sixties has set out to complete what his wife calls his "unfinished business." Nick himself makes a joke of the calling he chose late in life: "I was in foster care when I was five, and here I am almost sixty years later and I'm still in foster care."

In 1996, Nicholas Scoppetta became commissioner of New York's child-welfare agency, a $1.2 billion a year organization charged with protecting and caring for the precious victims of neglect, abuse, and the absence of love. The agency does environmental cleanup: when parents make a mess of it, social workers are expected to step in and provide a new, healthier environment in which to raise the children. Nicholas Scoppetta knows what it's like to be raised in hardship and how it feels to be taken away from your parents. He cannot speak of his tiny charges without his voice breaking.

Scoppetta rejected the life of a wealthy lawyer and the relative ease of a corner office on Madison Avenue to do a job that cannot succeed 100 percent of the time and in which a single failure means the death of an innocent child. A psychoanalyst might say that Scoppetta is working through his childhood problems, patching up the holes in his early life by giving something back to the world. That may be true, but the more interesting question is how did young Nicky beat the odds and triumph? How did a kid with none of the advantages in life rise to the top of a competitive profession and then be emotionally secure enough to give it all up to pursue a noble goal? Where did he get the smarts, the self discipline, and the ambition to succeed? How does anyone overcome such a challenging environment? Given the difficulties that so many children face, perhaps we should be amazed that so many of them turn out so well.

Asked to explain Scoppetta's success, a childhood friend named Ralph Pelligra, now a physician in California, said, "I tended to think of it as just native talents and native intelligence overcoming obstacles in the environment." Native talents and native intelligence. Innate. Inborn. Derived from the constitution of the mind, as opposed to being derived from experience. Genes.

We love stories like Scoppetta's because they celebrate

triumph over adversity, the victory of human will, and confirm our natural goodness. But seen another way, these stories show that maybe upbringing doesn't matter that much for intelligence, or for social skills, or success in life, that environment is not necessarily the most important part of the equation. What if a person's native intelligence and mental capacity are largely set from birth, no matter where and into what conditions that birth takes place? Perhaps it is time to look at where intelligence really comes from, instead of assuming that it's mostly the product of parents who talk to their babies, good schools, or summers at computer camp.

The roots of mental ability, of thinking and consciousness, are as complicated as the process of thinking itself. The brain is a physical organ, but it's not like other organs. A baby raised with a good diet and clean surroundings, but without human contact, will grow up with a set of perfectly normal body parts, except for one. Without human contact, the human brain will not develop. Nor is a brain an empty receptacle into which knowledge can be dumped or a blank computer into which data can be entered. It is more like a garden, a living ecosystem where all the parts are interdependent. The thinking, conscious brain is where nature and nurture are inseparable; one cannot exist without the other. The brain is built by biology, developed by human contact, and realized through social interaction. Its highest, most rarefied product is thought.

The brain begins with genes. Genes from both parents combine to design and create the lump of gray flesh in the head, plus the rest of the body, which operates and is operated by the brain. The slightest disruption in the physical development of the brain can have a devastating effect on future intelligence. A single glitch in the DNA code can limit mental development or cause severe retardation. On the other hand, "good" genes make genius possible. There is no single factor more important in an adult's IQ score than genes. However, parents also pass along the environment in which the genes are expressed. What we think about, the language we think in, and how we apply our intelligence are all products of the environment; how well we think depends very much on that original set of blueprints from the genes. Just like a person who hops into a car decides where to go and how to get there, the car sets limits on how fast and how successfully.

We spend more time thinking than in any other activity. It might be complicated thoughts about the meaning of life, or more likely whether to have a cup of coffee or go to the bathroom. Sometimes it is difficult to control thinking and to concentrate on a specific topic or to avoid thinking about a particular problem, but what is even harder is to stop thinking entirely, even for a moment. People spend years practicing meditation to attain an altered state of consciousness in which they are free of thoughts. Occasionally we lose a train of thought or say the mind went "blank," but quickly the thoughts pick up again and roar off in another direction. Too many thoughts can be distracting, and people with certain mental disorders often complain about "noise" in their brains. A common experience for everyone is lying in bed at night, endlessly replaying the day's events and trying to force sleep to come.

Thinking is also the most complicated of our activities. We categorize. We form concepts. We make plans. We reason. We make decisions. We hope, fear, remember, communicate. Psychologists have developed tests that separate these different aspects of cognition, but when it comes to how we actually think about real life problems, we usually use many or all of the different processes simultaneously. Each person's individual style of thinking—the unique intellectual style—is an intricate combination of mental facets brought together in the cerebral cortex, the largest and most modern part of the brain. For the most part human brains are similar to the brains of other animals, except for the cerebral cortex, which makes us unique. This part of the human brain is bigger and more complicated than in any other species, even among our closest primate relatives.

MEMORY

In many ways, we live in the past. Our thoughts are dominated by memories of what we have seen, heard, and experienced. Without the ability to remember we wouldn't know where we live or what we do. Nor could we understand this sentence. According to Eric Kandel, a neurobiologist at Columbia University's College of Physicians and Surgeons, memory is "who we are." Kandel is a pioneer in understanding the molecular biology of memory; his work has led to the surprising conclusion that memory works the same way in the simplest animals as it does in the most complex mammals like ourselves. More surprisingly, memory in the lowly sea slug and in highly developed mammals comes from exactly the same genes.

There are two basic types of memory: short-term and long-term. Short-term memory, also known as working memory, operates over seconds whereas long-term memory lasts from minutes to an entire lifetime. If you hear a random telephone number, 441–9620, you might remember it until the end of the sentence, but probably not until the end of the paragraph. Your own telephone number, however, is securely stored and easily remembered. The reason is that the random telephone number went into short-term memory only, because there was no reason to save it, while your own number is stored in a personal long-term vault.

Short-term memory is similar to the random access memory (RAM) on a computer; it controls the information needed to run at any given moment. Long-term memory is like the hard drive; a repository for all the information needed to operate. Just as a computer needs both RAM and a hard drive, so both short-term and long-term memory are essential for intelligence. For example, adding the numbers 349 and 217 requires long-term memory to remember the rules for addition, and short-term memory to execute the specific problem. Memory itself is so central to the thinking process that memory is one of the best predictors of human intelligence as measured by IQ tests.

How is information in short-term memory converted into long-term memory? It must be a selective process. Otherwise long-term memory would soon be swamped with useless information such as restaurant menus, road signs, and old *TV Guides*. It would be like the hard drive of a computer that stored every revision of every document, or a radio that recorded every song. The machinery would soon be filled up with useless, disorganized information. Somehow the brain must have a filter to sieve out what needs to be remembered from what can be discarded.

The filter is a physical structure built by genes. It was discovered by studying a simple invertebrate, the sea slug Aplysia. Sea slugs barely have a brain, and they probably couldn't pass the bar exam, but they do have a nervous system and are able to "remember" simple stimuli and respond accordingly. One of the best-studied responses is the gill-withdrawal reflex. When the gill of a sea slug is touched, the body withdraws into its shell; presumably the touch is a warning that a predator may be nearby. But if the gill is touched repeatedly, the withdrawal response slows down or disappears, as if the sea slug knows that it has nothing to fear. To the extent that intelligence is the ability to adapt behavior to the environment, the sea slug shows a primitive form of intelligence.

Eric Kandel wanted to know how the sea slug adapts its response. The first step was to recreate the reflex without the sea slug, using isolated nerve cells grown in a petri plate. By recording the electrical signals between nerve cells, Kandel found that after a single stimulus there is a strong electrical signal at the synapse between nerve cells, but as the stimulus is repeated, the strength of the synaptic connection decreases. The nerve cells are "remembering" their past. Kandel showed that the nerve cell "remembers" by synthesizing a burst of proteins and that the key activator of this explosion of gene expression is a protein called CREB. Kandel proved that the nerve cells could be fooled into thinking they'd been stimulated simply by adjusting the amount of active CREB protein.

The discovery was big news in the world of invertebrate neurobiology, but it was barely noticed by human psychologists. How could the response of a lowly sea slug—actually just a plate full of nerve cells—have anything to do with something as complex as human memory?

Tim Tully was pretty sure the finding was important. A young scientist at the Cold Spring Harbor Laboratory in Long Island, Tully studies the behavior of the geneticist's favorite pet, the fruit fly Drosophila melanogaster. Tully wanted to understand the way fruit flies remember how to behave in response to odors. In an experiment, he let the flies smell one substance and gave them a painful electrical shock. Then he gave them another substance to smell, without a shock. Pretty soon, typically after about 10 training sessions, the flies avoided the first substance but not the second. The flies were being "intelligent"; they were learning from experience.

Tully established three things: that the flies, just like humans, have two forms of memory, short-term and long-term; that short-term memory is required to learn the difference between odors, whereas long-term memory is required to remember the difference and behave accordingly; and that converting short-term memory into long-term memory requires new gene expression. But what genes were turning on? Taking a clue from Kandel's work on sea slugs, Tully decided to look at the CREB mechanism.

First, he genetically engineered flies that were low in CREB. He did this by giving them extra copies of the gene for the CREB repressor, which limits CREB production. Tully exposed these flies to the first odor and turned on the shock. With the second odor, no shock. He repeated the sequence again and again, but the flies didn't get it. No matter how many times they were trained, they couldn't remember to avoid the odor paired with a painful shock. Further tests showed that they could *learn* the difference between the "good" and "bad" odors; they just couldn't *remember* the difference, and so they kept getting shocked. These flies had lost the genetically controlled ability to have long-term memories.

Next, Tully produced a line of flies with extra CREB,

which he did by breeding them with an extra copy of the gene for the CREB activator, a molecule that increases the production of CREB. Tully then exposed them to the different odors and electrical jolts. These flies got it, and they got it incredibly fast. The high-CREB flies not only could remember the odors, they didn't need the typical ten training sessions. They learned in just one try. Tully marvelled that it was as if the insects had developed a "photographic memory."

For scientists, at least, it was amazing that the same genetic mechanism used for a simple reflex in sea slugs is conserved in the much more complex behavior of fruit flies. But would it apply to mammals? Discovering how the process worked in a mammal would be a major step up the food chain and would bring the research closer to humans. Ancino Silva, also at Cold Spring Harbor, decided to try the experiment on mice. Silva knew that normal mice find out what kind of food is safe to eat by smelling the breath of other mice. If a healthy mouse has corn on its breath, other mice know it's safe to eat corn. When Silva engineered a strain of mice that were low on CREB, they couldn't pass the smell test. In fact, they were virtually incapable of forming long-term memories.

The experiments showed the link between CREB and memory in sea slugs, flies, and mice. Although there is no proof yet that the same pathway is used by humans, our close genetic and biochemical similarity to mice makes it very likely. This does not mean that CREB by itself is responsible for making memories. CREB acts like a filter, but it is only one element in a long series of reactions.

It's possible, however, that the animal experiments already shed light on how we learn. Cells, animal and human, have only a limited amount of the CREB activator protein.

As a result, the amount of new information that nerve cells can deposit into the long-term memory bank is limited. This may explain why multiple short periods of learning are much more effective than one long period; the nerve cells need time to regenerate CREB activator. Silva showed this was particularly true for his forgetful mice; the only way they remembered anything was from multiple short training sessions with rests in between. Applying the same logic to human learning explains why all-night cramming for an exam does not work as well as studying a little bit every day, something that most students know even if they often ignore it.

MENTAL MAPS

You check into a strange hotel late at night, unpack your bag, shut the curtains tight, fall into bed exhausted, and soon are sound asleep. Suddenly a loud siren . . . FIRE! The room is pitch black, and yet you find your way to the door without a thought and head to the nearest exit automatically.

How did you do that? When you walked into the room the first time, without any conscious effort your brain made and stored a map of its new surroundings. Even as you were unpacking and thinking about other things, your brain was recording the layout of the room, the position of the furniture, the location of the door, and the direction to the nearest exit. The part of the brain responsible is the hippocampus, which makes a mental map in minutes and stores it for weeks. Later, if it is an important map, the information is transferred to the cerebral cortex for long-term storage. A damaged hippocampus, as can occur from injury or stroke, would prevent a person from finding his way out of a new room, even though he could remember the layout of a place he had lived long before.

The hippocampus records spatial information in large complex nerve cells called place cells. Each place cell contains information about one snippet of new territory. This was shown by exposing a mouse to new territory and recording the electrical firing of the place cells during a second look. Each time the mouse moved its head to see a different part of the territory a different set of place cells fired. Curiously, there is no correspondence between the position of the place cells and the information they store; they are spread around like pieces of an unassembled jigsaw puzzle. This is probably because the same place cell can retain information about different areas.

How did the information get from the eyes to the hippocampus? The key player appeared to be one form of a receptor that recognized glutamate, a small molecule used by brain cells to communicate. The theory was impossible to test, however, because mice bred without the glutamate receptor couldn't live. The glutamate receptor, like many brain proteins, was responsible for a variety of things, including maps, and mice without it could not survive. Finally, Nobel Prize winner Susumu Tonegawa at the Massachusetts Institute of Technology figured out what to do. He developed a clever strategy to specifically disable the glutamate receptor only in the hippocampus and not the rest of the body.[2]

Then the mice were given a map-making test. They were thrown into a pool of water that contained a barely submerged platform. The mice swam around until they found the platform, on which they could stand. Normal mice quickly learned the location of the platform. After a few training sessions, they formed a mental map of the landmarks and swam directly to the platform. But the mutant mice lacking the glutamate receptor in their hippocampus couldn't remember where the platform was. Even after multiple training sessions, they kept swimming blindly, as if they had never been in the pool before. Otherwise, the mice were quite normal and were perfectly good at other memory tasks that didn't involve remembering places. This showed that the glutamate receptor gene is both critical and specific for the type of thinking involved in making a mental map.

Even though the loss of a single gene can prevent a mouse from finding its way, the thought process depends upon more than just genes. A simple experiment shows that experience is important too. Some mice were raised either in a sparse, unfurnished cage with only a water bottle and food tray, while others grew up in a special "playground" equipped with plastic tubes, a tunnel with multiple openings, and an exercise wheel. After three months, the mice brought up in the more stimulating environment showed a 15 percent increase in the number of cells in the hippocampus. The more the mice used their brains to remember the complex topography of the playground, the better their brains became. Even for this simple type of intelligence, environment makes a difference.

MEASURING INTELLIGENCE

The discoveries of the role of specific genes in long-term memory and map making were possible because scientists were able to come up with tests that isolated those particular types of thinking. The key to the experiments was to be able to measure very specific and narrow facets of cognition that could be recognized in experimental animals. For years, however, the most common measure of thinking in humans has been for the most complicated function of the brain—general intelligence—and the measure of it the least specific—the IQ test.

Long before neurobiology, or even before science, people recognized individual differences in intelligence. We have always measured each other for intelligence, and we have informally created our own standards. A person can be street smart or bookish, lack common sense, be sharp as a tack, or dumb as a post. There are absent-minded professors and clever graduates of the school of hard knocks. Some people are good with words and bad at math; some can find their way in a strange city and others get lost in the supermarket. People are quick or slow, have good memories or are forgetful. A person can be wise as an owl or crafty as a fox.

Coming up with a reliable, quantifiable measure of intelligence has been much harder. In fact, there isn't much scientific agreement on a definition of intelligence. Lewis Terman, the pioneer in intelligence research during the 1920s, called intelligence "the ability to think abstractly." David Wechsler, who developed his own widely used intelligence test in 1944, defined it as "the capacity to understand the world and the resourcefulness to cope with its challenges." In 1982 Robert Sternber and William Salter called intelligence "a person's capacity for goal-directed adaptive behavior."

None of these definitions is encompassing enough to include all the different forms of intelligence, nor specific enough to indicate what is and is not included. For example, the ability to slam on the car brakes is clearly a goal-directed adaptive behavior, where the goal is to stay alive, but it's not what we normally think of as intelligence. Thus, psychologists have fallen back on a purely operational definition of intelligence: intelligence is what's measured by an IQ test.

Intelligence tests were invented because the French school system was too crowded. The year was 1905 and the French government had just made education compulsory for all children. As a result the classrooms were overflowing. Teachers were, in many cases for the first time, encountering children who weren't capable of keeping up with the standard curriculum. The government wanted to set up special classes to help these children, but no one was sure how to identify them objectively.

The answer came from Alfred Binet, the leading French psychologist of his era. Binet and his collaborator, Théophile Simon, devised a test to sort out children of average intelligence from those below normal, which became the predecessor of modern IQ tests. The test consisted of 30 questions in order of increasing difficulty. For example, an easy question was to point to the nose, eyes, and mouth; an intermediate question was to name four colors; and a difficult question was to make sense of a jumbled sentence. Binet reasoned that all children follow a similar path of intellectual development but some progress slower than others; i.e., they are "retarded." Thus, by comparing a child's test score to his or her chronological age, teachers could determine the child's ability to profit from the standard curriculum or from special education. The simple test did prove to be a reliable predictor of a child's success in the French school system.

A few years later, the German psychologist L. Wilhelm Stern formalized Binet's concept of mental age into the "intelligence quotient," or IQ, which is simply the ratio of mental age to chronological age multiplied by 100. The formula is designed to give an average IQ value of 100 for the population at large. For example, a five-year-old who performs as well as an average seven-year-old has an IQ of 140, while a ten-year-old who can only answer the same number of questions as most eight-year-olds has an IQ of 80. Because intellectual development does not proceed indefinitely, IQ tests for adults use broad age groups rather than exact chronological age. Since then many different IQ tests have been developed to measure aptitude rather than actual achievement or learning. The tests emphasize abstract thinking, such as comparing the areas of two geometric figures, rather than specific knowledge, such as knowing a particular equation.

It's clear that different people have different types of mental abilities. Bill might have a great vocabulary but be lousy at arithmetic, while Mary might be a math whiz but have a terrible memory. But it's also clear that people who are particularly bright tend to be good at many different types of mental activities, while less bright people rarely excel at any. IQ tests reflect this diversity of skills by asking different types of questions. When the scores from thousands of IQ tests are analyzed, two results emerge. First, there is a general cognitive ability factor, called the "g-factor" or simply g, which cuts across all the different types of intelligence measured by an IQ test. People who are high on g tend to do well on all aspects of the test, while those who are low on g do not fare as well across the board. Second, there are specific areas of mental expertise that are partially but not completely distinct, such as word fluency, numerical calculation, spatial visualization, memory, and so on. So overall intelligence depends on a combination of general ability (g) and specific expertise.

Although the significance and usefulness of IQ tests are passionately debated, one point should not be lost: IQ tests do what they were originally intended to do, they predict a person's ability to perform in school. In study after study, IQ is the single best predictor of school performance. Although many other factors, such as socioeconomic status and parental occupation play some role, there is no other better predictor of a child's grades or of how far he will proceed in his education. Nor should this be surprising since much of educational advancement depends on exactly the sort of skills measured by IQ tests. To a certain extent, IQ tests are simply measuring the ability to take tests.

Obviously IQ tests cannot measure the full range of human intelligence; it's hard to imagine any test that could. They also are culturally biased; even the smartest English speaker is going to fail an IQ test in Chinese. Cultural bias is not as evil as some critics suggest, however, because one measure of intelligence is the ability to understand and adapt to the environment, including culture. There have been criticisms that IQ tests are prejudicial and that it would be unfair to ask a city resident questions about farming or to quiz an American about the Chinese language. The well-designed tests, however, do not measure the accumulation of facts but the ability to think, and the tests themselves are constantly being evaluated and used on different cultures and in different languages. Nor are IQ tests a measure of a person's worth, but then they were never intended to be. Despite the objections and the limitations, the tests do measure something that relates to different types of mental ability, that has real-life consequences, and that differs in a cohesive way from one person to another.

THE GENETICS OF IQ

Why do different people, even when they are brought up the same way in the same environment, get different scores on IQ tests? Although there is no single answer to this question, the results of decades worth of studies on tens of thousands of subjects have been remarkably consistent in showing that the single most important factor is genes. The "environment" includes many factors that influence intelligence, such as prenatal care, nutrition, child care, schooling, etc. Together they are a powerful force, but not one of those environmental factors alone has a greater impact than genes.

Three different methods have been used to precisely measure the role of genes in IQ. Because the topic is so controversial, it's worthwhile looking at the numbers. The best method is to look at identical twins raised apart because they share all the same genes and none of the environment. An analysis of 158 such pairs gave a correlation of 0.75, which indicates that IQ has a heritability of 75 percent. That means that three-quarters of the differences between IQ scores are because of differences in genes.

A second method is to compare identical to fraternal twins raised together. If the correlation is greater for the identical twins than the fraternal twins—if identical twins are more similar for intelligence than fraternal twins—then some role of genes is implied. Looking at many different studies performed over seven decades on more than 10,000 twin pairs, the median correlations were 0.86 for the identical twins and 0.60 for the fraternal twins. Because these twins were raised in the same environment, the best way to estimate heritability is by doubling the difference between the identical and fraternal pairs, which gives a value of 52 percent.

The third approach is to look at adopted children and their parents and siblings. The correlation between 720 pairs of biological parents and their natural children was 0.24, even though the children were raised entirely by someone else. The correlation between 203 pairs of genetic siblings raised by different adoptive parents was also 0.24. Because these are first-degree relatives who are only 50 percent genetically related, these correlations must be doubled, which gives a heritability estimate of 48 percent.

Thus three quite different experimental designs all lead to the same conclusion: IQ test scores are substantially heritable. This is true regardless of the way IQ is measured, who is measured, or when it is measured. One can quibble over whether the "real" figure for heritability is 48 percent or 75 percent; combining the total world's literature on twin, family, and adoption studies gives an estimate of about 50 percent. What one can't argue, at least based on the numbers, is that heritability is nil. No matter in what country, in which historical time period, in what age group, or with what type of test, the result is always the same: no other single factor is more important than genes in determining cognitive ability. What would be remarkable at this point would be a scientific study in which heredity was *not* important for IQ.

IQ AND THE ENVIRONMENT

Just because genes are important for IQ doesn't mean that environment isn't. In fact, the same data that show IQ is heritable also show that shared environmental factors play an important role. The most powerful method of looking at environmental impact is to examine genetically unrelated siblings adopted together. The only thing they share is the environment, so any similarities must come from the way they were raised. The correlation for this type of pair is 0.32, which means that factors such as being raised by the same parents, attending the same schools, and growing up in the same neighborhood can make genetically unrelated children raised together 32 percent more similar than if they were raised apart. Similarly, when parents who adopt are compared with their nonbiological adoptees, the correlation is 0.19. Again the similarity, in this case 19 percent, must come from common environmental factors rather than from genes. Finally, recall that IQ scores are correlated 0.86 in identical twins raised together compared with 0.75 in twins separated at birth. The difference indicates that about 11 percent of the variance in IQ is from those environmental factors shared by twins in the same household.

These three different comparisons show that environment, in particular the type of things shared growing up, is important for how well one scores on an IQ test. Again, the number could be anywhere from 11 percent to 32 percent, but the important point is that it's greater than zero.

Ironically, it is these experiments designed and carried out by behavioral geneticists that provide the best evidence for the environment's role in IQ. When social-minded psychologists have tried to measure the effects of the environment without controlling for genetics, such as by comparing test scores with the number of books per household, they have never been able to conclusively prove that environment plays any role at all.

The experiments do not, however, show *what* about the environment matters. Some of the best evidence comes, sadly, from orphanages where children received little human contact or stimulation. The classic experiment of this type was started in the 1930s by H. M. Skeels, who studied children in an Iowa orphanage who had been classified as mentally retarded and were receiving almost no human attention. Thirteen of these children were moved to a separate institution and placed in the care of foster mothers who spent much of their time nurturing the children. Within just four years these children had gained an average of 30 IQ points, a remarkable increase, whereas 12 other children who remained in the orphanage had lost 20 IQ points. Twenty years later the differences remained. Most of the children who received individual care graduated from high school and became self-sufficient, whereas most of those who grew up in the orphanage were either still institutionalized or not self-supporting. What's really amazing is that the children who were moved were placed in an institution for mentally retarded adults; the foster mothers were themselves mentally retarded. So, it's not necessary to be a genius to help a child's mental development; what is important is love and human contact.

Similar results have been described in an orphanage in Iran where kids thought to be mentally retarded were given individual care and improved markedly in language skills. In a French study, children adopted into homes with high socioeconomic levels increased IQ scores by an average of 12 points. In the United States, the "experiment" is carried out every day in Project Head Start, which provides children from disadvantaged homes with one to two years of help with intellectual and social skills beginning at age four. The program reports significant short-term gains in IQ test scores and some long-term improvements in school performance, social skills, and self-esteem.

The gains fall off as the children grow older, however, and some experts now suggest that starting with four-year-olds is too late. University of Alabama scientist Craig Ramey has designed a program for the children of retarded mothers that starts six weeks after birth and lasts five days a week for a year. The babies get intensive one-on-one contact with a trained teacher, lots of talking and responding to the child, and plenty of hugs and kisses. Ramey finds that the program has reduced mental retardation in this genetically vulnerable group by as much as 50 percent. However, he warns that parents should not get carried away and overstimulate babies to the point where they are exhausted or fed up; like adults, babies need quiet time.

While the most important aspect of nurturing for IQ is probably the right amount of human contact and stimulation, there are many other factors that matter. One of the most obvious, but also overlooked, is proper nourishment, both for the pregnant mother and for her child. The brain, like the rest of the body, needs food to properly develop, not only a balanced diet but also one free of toxins. The devastating impact on IQ of just one pollutant found in Great Lakes fish was documented by Joseph and Sandra Jacobson of Wayne State University in Detroit. The Jacobsons looked at 212 children who were recruited as newborns. At the time of birth the mothers' blood and breast milk were measured for a once-common industrial pollutant known as PCB. When the children were tested at age 11, the IQs for those with the highest prenatal PCB exposures were more than six points lower than for those with smaller exposures. The biggest effects were on memory, attention, and planning skills. The exposed children also lagged behind in reading skills.

The thing to remember is that no matter how strong the genetic potential for IQ, a slight environmental disruption can be devastating. The wrong foods, too many toxins, or insufficient stimulation can all lower IQ scores. Since you have little say in what genes you give to your child, except in your choice of mate, the only alternative is to focus on the environment. There are countless guidebooks about how to raise bright babies and some disagreement about how to do it. The secret seems to be to engage the child fully as soon as possible. Active, supportive, unconditional love is probably the greatest gift.

RACE AND IQ

The connection between genes, environment, and IQ has become enmeshed in one of the bitterest scientific feuds of our times: the role of race in intelligence. In the United States, Asians as a group score about 3 IQ points higher than whites, who in turn score about 15 points higher on average than blacks. What's controversial is not that such group differences exist, but what they mean.

First of all, it's important to emphasize that these are only average differences. There is a great deal of overlap between the groups, and all the races show the same broad range of IQ scores. For example there are many Asians who score lower than the average white or black, and many African Americans who score higher than most whites and Asians. This means that knowing a person's race is not a good predictor of his or her IQ. There are also regional differences in IQ scores, and for that matter in ability to play basketball—but that doesn't mean you can judge how smart a person is, or how good a free-throw shooter, by asking what state he's from.

Nevertheless, there is still the puzzling issue of a substantial point spread in average IQ values. One theory is that the difference is genetic. This theory was most recently brought to national attention, and intense emotional reaction, by the 1994 book *The Bell Curve* by Charles Murray and Richard Herrnstein. They began with two facts: individual differences in IQ scores are substantially heritable and race is heritable. From there, they deduced that racial differences in IQ scores must also be genetic.

Their argument is fundamentally unsound because it confuses individual differences, which is what twin studies and other genetic methods measure, and group differences, which can't be addressed by these methods. Consider an example in plants. Suppose a bed of sunflowers is sown carefully in a sunny spot and diligently tended and watered. All things being equal in this ideal environment, the height that each sunflower grows would depend predominantly on its individual genetic makeup. The quality of seed is the only variable that matters. But suppose we carelessly dumped half the seeds in a cramped and partially shaded bed and forgot to water them. Depending on where they fell, some of the "good" seeds would not do as well as the "bad" seeds. If the average heights of the flowers from the two beds were compared, the differences would depend entirely on the environment. The only conclusion that could be deduced from the data would be that the gardener should be fired.

The real world is like that garden; the variable quality of

the environment is so great that it's hard to say anything about the seeds. We simply do not know whether the group or racial differences in IQ scores have any genetic basis or if they are like the heights of sunflowers grown under different conditions. On the other hand, we do know that IQ is at least partly determined by environment and that some of the environmental factors known to be important are precisely those that vary according to race in our society. For example, it's known that IQ can be influenced by education and that Asian Americans on average place a greater value than other groups on education, scholastic diligence, and achievement. IQ is also correlated with socioeconomic status, which in the United States remains lower for those of African than of European descent.

Direct evidence for the importance of environment to racial differences in IQ comes from an adoption experiment described by Sandra Scarr and Richard Weinberg in the 1970s. They studied 99 African American children who were adopted from poor black households into white middle-class households in Minneapolis. The average IQ of these adopted children was 106, which was higher not only than the average black IQ but also than the average white IQ. The researchers estimated that being brought up in a relatively privileged environment contributed about 16 points to the IQ scores— just about the size of the difference between average IQ scores for whites and blacks in the United States.

Racial mythology is such a powerful force that it may itself push down IQ scores. Claude Steele and Joshua Aronson administered a difficult verbal test to black and white students at Stanford University. They told some of the subjects that they were taking a test of intellectual ability, thus presenting the black students with the threat of fulfilling the stereotype about race and intelligence. The other students were told that the test was simply a laboratory problem-solving task that had nothing to do with intelligence, thus making the racial stereotype irrelevant. For the white students it made no difference what type of test they thought they were taking. But for the black students, merely the idea that they were taking an intelligence test lowered their scores by more than 25 percent—a margin even larger than the black-white difference in IQ scores. As Steele and Aronson comment, "Compared to viewing the problem of Black underachievement as rooted in something about the group or its societal conditions, this analysis uncovers a social psychological predicament of race, rife in the standardized testing condition, that is amenable to change . . ."

MAKING BRIGHTER BABIES

Part of the reason that *The Bell Curve* was so controversial is that if IQ is genetically fixed at birth, why should society bother with Head Start or other programs to help black children? If they aren't going to get smarter, why throw good money at bad genes?

This is another example of not understanding how genes work. A person's genes are entirely in place at conception, but they aren't all switched on. Humans are designed to develop slowly and cannot do so without the help of other people. Fetal development follows an orderly, predictable path, as does childhood. Milestones such as puberty must occur precisely on time or the child does not develop properly. Different genes are constantly dimming and brightening, switching on and off, and responding to a changing environment. The genes that control the development of the brain, and thus IQ, are no exception. This is why the power of genes to shape IQ changes over time.

Ronald Wilson and colleagues tracked several hundred pairs of twins in Louisville, Kentucky, for 20 years. They tested their mental development and IQ starting at three months and continuing at regular intervals up to 15 years of age. In infancy, the correlations for the twins were quite high and showed no difference between identical or fraternal twins. At this very early stage, the environment provided by the parents was paramount and genes were less important. Then an interesting change began. As the children got older, the correlations for the identical twins increased, whereas those for the fraternal twins decreased. The genetic similarities of the identical twins were making them more alike in intelligence, while the genetic differences of the fraternal twins made them more dissimilar. By age 15, the heritability of IQ had increased from almost nothing to 50 percent or 60 percent, which is similar to that seen in adults.

The genetic influence on IQ increases with age through adulthood. Matthew McGue and colleagues have examined data from a large number of twin studies involving subjects between the ages of 11 and 88. As the twins progressed from adolescence to middle and older age, the heritability of IQ increased from about 50 percent to as high as 80 percent. Meanwhile the effect of the shared environment dropped from being very important to close to zero, which might be expected as the twins moved away from home, parents, and shared schools.

Thus, IQ develops dynamically, not statically. Environment is most important early on, while genes become most important as we mature. This finding is reenforced by the observation that the extent to which a child's future IQ score can be predicted from its present score increases over the same period. A likely interpretation is that the environment for infants and young children is almost completely provided by their family, while as people mature they become more responsible for creating their own environment. For example, how much a child is read to is determined by its parents, but how much an adult reads depends only on himself or herself. There is an early window when much can be done to maximize IQ, but that window closes with time. To make a difference in a child's IQ, start early.

THE SEARCH FOR IQ GENES

The evidence is strong that IQ is largely determined by genes, but that's a long way from identifying the specific genes. Typically, the best indicators of what effects the genes might have come from studying the extremes, in this case the low end of intelligence. There are single genes known to cause mental retardation, a single stutter in the language of DNA that will prevent a person from reaching normal levels of intelligence. It's not yet clear whether the same genes that cause retardation also influence the normal ranges of intelligence, but they do offer a glimpse into how complicated the search will be.

Intelligence is similar to height in that there is no fixed point that defines "tall" or "genius" or "short" or "retarded." The often used definition of retarded is an adult with an IQ score less than 70 who is incapable of living alone. Such mental retardation is quite common; two to three Americans out of every 100 are mentally retarded by this definition. About 40 percent of cases are of unknown origin, and about 20 percent are because of known environmental factors such as lack of oxygen during birth and viral infections of the brain. The remaining 40 percent of cases are genetic. That means that 1 out of 100 Americans can't lead an independent existence because a genetic defect has stunted their mental ability.

More than 100 different mutations have been identified that can cause mental retardation. Many of these aren't necessarily found in brain-specific genes, but are in housekeeping genes that carry out the humdrum everyday business of cellular biochemistry. One of the best-studied examples is phenylketonuria, or PKU, which is caused by a mutation in the gene for phenylalanine hydroxylase, a liver enzyme that converts the amino acid phenylalanine to tyrosine. When the enzyme gene is broken, there is a buildup of toxic by-products. Since even one copy of the normal gene makes enough enzyme to avoid the buildup of toxic products, both copies must be mutated to cause the disease. Infants born with two copies of the PKU mutation appear normal at birth. However, as the toxic byproducts build up, producing a "mousy" smell in the urine, they cause irreversible damage to the brain and nervous system. The result is profound retardation and microcephaly. Untreated children never develop mentally past the infant stage.

Fortunately there is a cure for PKU, and it has nothing to do with gene therapy. The cure is simply to eat foods low in phenylalanine. So long as food is controlled from within a few weeks of birth, permanent damage to the brain is prevented. When the child becomes a teenager, the brain is developed enough to resist phenylalanine poisoning, and the special diet can end. An excess of phenylalanine can be spotted in a tiny drop of blood from a heel prick, and today every child born in most of the developed world is tested for PKU. Before the screening began in the 1960s, PKU accounted for 1 percent of severe mental retardation. Now it is virtually eliminated.

Another single gene cause of mental retardation is less easily cured. Fragile X syndrome is the most common form of inherited mental retardation, affecting 1 out of 1,250 live births. It causes mild to severe mental retardation and some physical abnormalities such as large ears and testes. This form of mental retardation was recognized as being linked to the X chromosome because it shows the usual pattern of being more common in males than in females. The strange thing was that it appeared to be getting worse as it passed from one generation to the next. For example, there were cases of a father carrying the mutated X chromosome who was perfectly normal, yet his child became severely retarded.

The first clue to what was happening came from blood cells. In patients with the syndrome, the X chromosome looked strange, so fragile that the very tip often was missing, as if it had broken off. The culprit was a trinucleotide repeat, a stretch of three DNA building blocks repeated over and over. Normal people have fewer than 50 copies of the repeat, but people with the disease have hundreds or thousands of copies, which leads to an inactivation of the gene and the fragility of the chromosome. Once the repeat hits a threshold of about 200 copies, it can only get bigger and not smaller, which explains why the disease gets worse from one generation to the next.

Surprisingly, the fragile X gene is not really about the brain. It produces something that is found all over the body, and the number of repeats has no effect on normal intelligence. Only when the number of repeats hits the danger point is intelligence harmed, and then the damage is dramatic. Just like with the PKU gene, the fragile X gene is working on a whole system of which the brain is a part. These genes are not acting on intelligence per se, but normal intelligence is impossible unless the genes work properly. With so many things like this that could go wrong during development, the amazing thing is how smart we are.

LANGUAGE GENES

"It's a flying finches, they are."

"She remembered when she hurts herself the other day."

"The boys eat four cookie."

"Carol is cry in church."

These confused sentences were uttered by American adults whose native language is English. The speakers, all members of the same family, share what is known as a specific language impairment (SLI), in this case a problem with verb forms. The family, discovered by linguist Myna Gopnik and described in Steven Pinker's book *The Language Instinct,* didn't pick up this manner of speaking at school or learn it at home. What is mangling their language is most likely a genetic defect.

There is growing evidence that genes play an important role in language. One line of evidence comes from the fact that language is so specific to our species. Just as humans have the greatest ability of all the species to think, so they have the greatest ability to communicate thoughts. While Lassie and Flipper may be able to communicate some pretty complicated stuff, there is no animal parallel to our complex language, language that can distinguish past, present, and future and express abstract concepts. There must be something about the human genetic blueprint that makes us especially capable of language.

Another line of evidence points to the complicated rules found in all languages. Every language has its own precise grammar, syntax, spelling, and pronunciation. These rules can be a nightmare for an adult trying to master a foreign language, yet small children are perfectly capable of speaking their native language like, well, natives. This is what originally prompted linguist Noam Chomsky to theorize that a newborn's brain is prewired with the ability to recognize sounds and to learn the basic rules of grammar and syntax. He postulated that the human brain comes preequipped with an "innate language acquisition device." According to his theory, the specific language a child learns depends completely on the environment, but the basic ability to learn the rules and apply them to the billions of different combinations is genetically preprogrammed.

One way to isolate the specific genes may be to study what happens when the "language acquisition device" goes awry, such as in the family described above. In such cases, children start talking late, and then they have problems articulating words. They make grammatical errors, which may persist through adulthood. They can have perfectly normal intelligence and be skilled in many things, except language. Specific language impairment runs in families, suggesting the possibility of a genetic basis. In the family described by Gopnik and Pinker, the grandmother has the impairment, as do 4 of her 5 children; they in turn have 23 children, of whom 11 have the language problem. However, one of the grandmother's daughters speaks normally, and so do all her children—as if this side of the family had escaped the gene. This is just the sort of pattern expected for a condition caused by a single dominant gene, although the gene itself has not yet been identified.

More progress has been made in finding the genetic roots for another language-related disorder, dyslexia. Once called "word blindness," dyslexia is characterized by difficulty in learning to read despite adequate intelligence and education. It seems to be caused by a fundamental brain disconnect between written words and their meaning. The malfunction is specific for reading; dyslexics can be very bright, even brilliant in other areas; they "just" can't read. As many as 8 percent of children are reading disabled by standard school system criteria. Children with dyslexia are the largest group of students who receive special education services.

Dyslexia was first recognized as a discrete disorder in 1896, and within a decade it was known to run in families. Since then many family and twin studies have shown a substantial genetic effect. Most dyslexics have at least one other person in the family who has reading problems, and the rate in identical twins is as high as 40 percent.

Even though a reading disability is clearly a complex disorder with many different causes, both inherited and environmental, there appear to be at least some families in which one major gene is involved. This led scientists at the University of Colorado in Boulder and at the Boys Town National Research Hospital to start a gene hunt in families with at least two dyslexics. They found evidence linking reading scores to chromosome 6 and found similar results in fraternal twins. More recently a different group of scientists, at Yale University, found linkage to the same chromosome in yet another group of dyslexic families. These look like solid results, although it's not yet known what gene is involved or how it works. One clue is that the Yale study found that the suspect piece of chromosome 6 does not affect all facets of dyslexia equally. The gene has a major effect on the ability to break down long words into syllables, but not on comprehension of short words. That suggests something quite specific—a breakdown in the brain circuit used to segment words—rather than a more general problem. So when the actual gene is found it may tell us something fundamental about that particular part of intelligence dealing with language and reading.

GENETIC INTELLIGENCE

Finding genes for retardation, language disorders, or reading problems likely will be easier than finding genes for overall intelligence in average people. Robert Plomin, one of the leaders in the field of behavioral genetics, has begun an ambitious project to collect DNA samples from a large number of children, ranging from genius to somewhat below average. He plans to test the samples with a series of genes likely to have something to do with neural functioning. Although Plomin's experiment is well designed, its success will depend on something that isn't yet known: the genetic architecture of IQ. While we know that IQ is substantially heritable, we don't know how many genes are involved. For example, if IQ is 50 percent heritable and there are 10 major genes, then each gene would have a 5 percent effect, not huge, but detectable using a few hundred subjects. But if there were 100 or 1,000 genes, then each would account for only 0.5 to 0.05

percent of the difference from one person to the next, an effect so small that it would be difficult to detect and understand even using thousands of subjects.

Most likely the genetic architecture of intelligence is much more complex than other parts of personality. One clue is the large number of genes that have been identified for mental retardation. It seems that almost any change in basic cellular metabolism can impinge on cognitive functioning. It would be surprising if there weren't at least an equal number of genes involved in the normal range of intellectual functioning. A second hint comes from evolutionary neurobiology. The brain's cerebral cortex, the complex thinking apparatus, is a recent addition compared with the limbic system, the seat of our emotions and temperament. The limbic system is like an old-fashioned plow: simple, sturdy, rugged. The cerebral cortex is more like a modern combine: complicated, high maintenance, easy to break. Deciphering the genetics of temperament and the limbic system may be accomplished with a rough sketch, but understanding the genetics of intelligence will require a far more sophisticated blueprint.

SEVEN

HUNGER

Body Weight and Eating Habits

I saw few die of Hunger, of Eating, 100,000.
—BENJAMIN FRANKLIN,
POOR RICHARD'S ALMANACK

Sandra was a beautiful baby, plump and giggly, with red cheeks and sandy hair. She rarely cried, soon learned to sleep through the night, and ate well from the day she came home from the hospital. She took to the breast right away and drank deeply and contentedly until she fell asleep in her mother's arms. At her first checkup, her parents were proud to learn that she was perfectly healthy and in the 80th percentile for height and weight. "Already she's at the top of her class," joked Sandra's father.

Sandra's mother struggled to regain her figure after the birth. She had spent her entire life trying to stay thin, or at least not to inflate like a blimp, which is what would happen if she didn't watch herself. When she was a girl, her own mother spoiled her with food. If she cleaned her plate, she got a "second dessert," which was a cookie or piece of cake, after her "first dessert" of fruit. Sandra's mother was still tempted and tormented by sweets. Now that she had a daughter of her own, she vowed not to repeat the error. In fact, Sandra's mother was so strict that she didn't allow Sandra to have *any* sweets until she was four years old. There were special occasions, of course, and if Sandra behaved well all day, her mother occasionally gave her a special treat of low-fat ice cream.

Like all kids, Sandra managed to find sweets and other goodies on her own, whether her parents liked it or not. When Sandra was five, she announced to her father, "Daddy, I love candy."

"Well, I love candy, too," said Dad.

"No you don't understand," the little girl insisted. "I *need* candy."

As a girl, Sandra was chubby, but not really fat. She had such a cute personality that people didn't seem to notice her size. Her mother called it baby fat and figured Sandra would slim down as she got taller. The truth is, she kept getting bigger. By the time she was 13, she outweighed her friends by 30 to 40 pounds. She had to buy clothing in unusual sizes, either large child sizes or small adult. By the time Sandra was a teenager, her weight was hurting her social life. While her friends started dating in high school, Sandra didn't really have a boyfriend. All the guys thought she was great, really fun and nice, but they didn't think about her as a girlfriend. When a big group of them went skating one winter, the boys skated away from her, yelling, "Don't go near Sandra, she might crack the ice!"

Sandra started her first diet at 16, the day after a particularly awful date. The diet wasn't anything extreme, just stuff she picked up from her mother, girls at school, and magazines. She ate her hamburgers without the buns, lots of cottage cheese and fruit plates, and she skipped dessert, except for the occasional vanilla milk shake as a special treat. The diet worked: she lost 12 pounds and felt fantastic. She loved how her old clothes looked so baggy, and to celebrate, she and her mom went right out and shopped for a new wardrobe. Steven, a boy at school who was president of the glee club and pretty popular, invited her to the prom, and she had a wonderful time. She framed a picture of herself from that night, wearing a pink chiffon gown and a corsage from Steven. She looked radiant and slim.

Sandra and Steven were married three years later. For her wedding, Sandra was determined to wear the same size dress she wore to the prom, but she needed a crash diet to make it. There was a lot of stress with the wedding preparations, but she kept to her diet and was able to squeeze into the dress on the big day. It was a long time after the wedding before she could eat another grapefruit. Steven was crazy about her, and he couldn't keep his hands off his new bride. She was soft in all the right places, and every night they made love before he snuggled against her to fall asleep.

Their first child, Steven, Jr., or Stevie, was born only a year after the wedding. Sandra lost most of the weight she gained during the pregnancy, but didn't want to get too thin because the baby was breast-feeding. When the second child, Christie, was born soon after, the pounds were harder to shed. There was no way she could do a grapefruit diet anymore, not with Christie breast-feeding, but Sandra did try to cut back on what she ate. The weight wouldn't come off, though, even when she knew she was taking in half as many calories as before. She tried an aerobics class for new moms, but it was such a hassle to get a baby-sitter. Anyway, she figured that running after two kids was exercise enough.

"Why should I pay to take classes, when I've got Stevie and Christie to keep me in shape?" she thought. Plus, her husband was so happy with his new family that he said he didn't mind the few extra pounds.

Their blissful life started to fall apart when Sandra was 35. It wasn't a single calamity; just a series of things that seemed to go wrong one after another. First Sandra's dad, whom she always adored, was diagnosed with prostate cancer, her mother fell ill, and Stevie went through a rebellious stage and was always fighting with his father. They didn't have enough money, and there was never enough time.

The stress caught up with Sandra and she came down with an awful case of the flu. She hadn't seen a doctor in years, and the clinic had some difficulty finding her old medical records. At first, the doctor thought there might be some mistake, because the woman in the file weighed 118 pounds. The woman breathing heavily and sweating big drops onto the floor was close to 160 pounds. She wasn't plump, she wasn't chubby. Sandra was obese. The doctor went to check her blood pressure and struggled to get the gauge around her beefy upper arm. After the checkup, the doctor told her that her pressure was up and she needed to lose a few pounds. "Oh, I've been under so much stress lately," she said. "I just need to get back to my routine."

The flu wiped out her appetite, so Sandra decided to make the most of it and begin right away on a diet. She drank gallons of water, which also was good for the flu, and ate only cabbage soup for two weeks. She lost eight pounds and thought, "Hey, this is easy." But as soon as she started eating regular food again, the weight returned.

Sandra didn't have time to worry about her weight, though, because things at home went from bad to worse. Her father had surgery for his cancer, but it was discovered too late. He died after spending three months in the hospital, just long enough to eat up all of his savings. Her mother needed to be in a nursing home, but there was no money left, so Sandra took her in. The old woman couldn't do anything for herself, but she never stopped criticizing everybody else, especially Sandra.

To pay the bills, Steve took a second job with a vending company, stocking the machines at the university. The car was always packed with candy, cookies, and bags of chips, and Sandra couldn't resist. She especially liked the Snickers bars when they froze solid in the cold garage. Steve was home far less often, and then he crashed in front of the television. He had very little patience for young Steve, who had grown up to be as wiry and high strung as his dad and spent his life under the hood of the car. Christie was doing okay in school but seemed way too concerned about boys and about being popular. She was practicing to be a cheerleader and was obsessed with her figure. She, too, had discovered the candy in the car and would sneak out for midnight snacks.

At 43, Sandra was exhausted. No matter how early she went to bed, she woke up tired. Her body ached, her hands and feet were always cold. With much urging from Christie, she finally made an appointment to see the doctor. He listened to her complaints, noted that her weight had climbed to 180 pounds, and ordered a series of tests. When the results came back, the doctor said she had adult-onset diabetes. "Well that explains my weight," she thought, somewhat relieved that there was a medical reason for being so heavy. She took the insulin shots, but there was no way to follow the strict diet and keep her family fed.

For Christmas that year, Christie bought a two-for-one

membership at a health club. She and her mom were going to start taking classes and watching their diets. Sandra went gamely the first time, but when she looked in the mirror, she saw a whale in a leotard and became extremely discouraged. When the instructor told them to touch their toes, Sandra couldn't; she couldn't even see her toes. That was the first and the last time she went to the health club.

From that point, Sandra's health deteriorated quickly. She spent entire days in bed, too tired to get up. A few months later, Sandra's kidneys failed. She died one gray morning, propped up on a pile of pillows. She was 52.

The death certificate read, "Renal failure secondary to diabetes." The official looking document didn't note the real cause of death: obesity.

At the funeral, one of Sandra's best friends and her husband stood before the ornately carved wooden coffin. "What a beautiful coffin," the woman said. Her husband whispered into her ear, "Yeah . . . must have cost a fortune for one that big." Christie overheard the remark and winced. What a pair of creeps, she thought. After the service, Christie retreated to the kitchen, which had been filled with casseroles, cakes, and pies donated by relatives and friends. She sat down in front of a very large German chocolate cake and started to eat, washing down each bite with a sip of Diet Coke.

Sandra's case is not unique, and obesity has become one of the most common health problems in the United States. The National Center for Health Statistics estimates that one of every three adult Americans over 25 is obese, meaning they weigh more than 20 percent above their ideal body weight.

Surprisingly—given the current emphasis on thinness, the development of sugar-free and fat-free foods, and the popularity of exercise—rates of obesity are increasing. The current rate of about 33 percent is well up from the rate of 25 percent in 1980. This increase is unlikely to stop anytime soon, since 21 percent of teenagers are already obese. The only thing growing faster than the national waistline is the bottom line of the weight-control industry. Americans spend between $30 billion to $50 billion a year on diet foods, diet books, exercise videos, weight-loss seminars, and support groups. Seventy percent of young women and girls age 14 to 21 are on a diet.

Unneeded body fat has both medical and emotional costs. Obesity is a major cause of high blood pressure, coronary heart disease, and some cancers. It's linked to adult diabetes, which occurs in 8 to 10 percent of white Americans. The problems are especially severe for women. The Nurse's Health Study found that women who were 20 percent or more overweight were twice as likely to die prematurely as other women. By contrast women who weighed 15 percent less than the national average had significantly lowered rates of early death. Every year, obesity and related conditions account for 300,000 deaths in the United States and are regarded as the second leading preventable cause of death after smoking. If obesity were an infectious disease, like tuberculosis or AIDS, it would be declared a national emergency and would become the target of a medical "war." Instead, doctors treat the various diseases caused by obesity but not the underlying cause. The medical establishment shares the general public misconception that being overweight is simply a matter of sloth versus will power—that fat people are that way because they're lazy and can't keep their faces out of the cookie jar.

Nothing could be further from the truth. Eating is one of the most ancient and evolutionarily conserved of our behaviors. If we didn't eat we'd die. More precisely, if we didn't eat enough to replenish the calories we burn, we couldn't survive. So naturally we have inherited genes that make our brains say "Eat!" We have evolved other genes that make our bodies conserve calories instead of burn them. Unfortunately, even though our genes haven't changed in the last several hundred thousand years, our culture has, dramatically. Not too long ago, at least in terms of evolution, food was rare and difficult to secure; most people probably spent much of their time trying to find it. Genes that encouraged people to eat and allowed them to store fat were helpful. But today, when food is plentiful in the developed world and more and more people spend their time at a desk instead of hunting, those genes are less essential. In fact, the same genes that saved lives a few thousand years ago can now be deadly.

Even during the past century there have been major changes in our environment. For example, the average American today has a choice of 50,000 different food products in the supermarket compared with about 500 products at the turn of the century, and the average American today expends only about one-fourth the calories of an average American in 1900. The current increase in obesity has nothing to do with genes and everything to do with how we live. The world we have created is radically different from the world we were designed for.

While our ancestors also were obsessed with food—finding it, that is—our obsession is more refined. Americans eat at least three meals a day, even if they aren't hungry, and munching, nibbling, and grazing occur from the morning coffee break until the final midnight snack. We "do" lunch and have romance with dinner. What would Sunday be without brunch? Or a ball game without beer and hot dogs? We patronize places that sell only certain types of food, such as hamburgers or tacos. We support establishments that serve nothing but hot water flavored with roasted beans. In urban America, we not only have coffeehouses but *competing* coffeehouses selling highly specialized varieties of flavored hot water. We have restaurants for all tastes and incomes, we shop for genetically engineered foodstuffs in climate-controlled warehouses, and we spend billions on advertising for competing brands of nutritionally worthless junk.

We should be grateful for the cornucopia of abundance, but we also must be careful. We have separated food from hunger at a time when the supply is unlimited. We satisfy other "hungers," emotional needs and desires, with food meant to satisfy physical hunger. An ice cream cone is offered as a way to cheer up, Coke adds life, and nobody can eat just one potato chip. We have refined techniques to increase the caloric content of food, loading it with sugar and fat, and also salt. Ironically, one of our most calorie-packed inventions, chocolate, is offered as a gift to the people we love, often in boxes shaped like hearts.

Clearly we have created an environment of great possibility and great risk. People in developed countries have benefited from improvements in food production and wealth, but we also have set traps that some of us will fall into. Some people are at greater health risk in this environment of abundance than they would be in a subsistence economy. Not too many years ago, the people at greatest risk were those who burned calories quickly and could not store fat. Those who survived in a harsh environment where food was scarce were able to conserve what they consumed. Today, the most desirable bodies belong to the ones who burn calories the quickest, to the people who can "eat anything" and not gain a pound. Those most at risk, and those least satisfied with their appearance, are the ones who store fat well.

FAT GENES

People like Sandra, her mother, and her daughter were born at risk in this culture of plenty. All three gained weight easily, burned calories with difficulty, could not control their diets, and suffered the emotional and physical consequences. People used to blame this kind of obesity on "glands." Then it was popular to believe body weight was controlled by "will power." Now most people recognize that body weight responds both to biology and society, to genes and emotions. The most recent experiments show that genes are the single most important contributor to body weight, more than any other factor or combination of things. The best estimate is that body weight is about 70 percent inherited. That means "will power" still has a role, but it takes an extraordinary effort to change the body's original design.

The evidence for the role of genes comes from the study of identical twins measured for their body mass index, which is weight corrected for height. One study of 93 pairs of identical twins who were raised in different households found the body mass index was 70 percent correlated. In other words, if one twin was fat, the other had a good chance of also being fat. Interestingly, the part that was not heritable was attributed to "unique" environmental factors, which means that shared childhood experience had essentially no effect on adult body weight.

This would seem to suggest that our efforts to watch our diets and instill good eating habits in our children are a waste of time. To check that surprising conclusion, scientists looked at adoptees, who shared a rearing environment but no genes with their new families. The results showed that the adoptees did have similar body-mass indexes with their families, but only with their biological families. The fat kids were likely to have fat biological parents, even if they had never met. By contrast, there was no relationship between the weight of the adoptees and the people who adopted them. It didn't matter if kids were adopted by vegetarian health nuts or donut dunkers; they grew up to be the size of their biological parents. One study even found that kids adopted into fat families grew up to be skinnier than normal.

But what if your dad has a beer belly and your mom is enormous? Or what if you look like your grandmother, who had thick hips and a skinny chest? Are you doomed to have their weight problems or wear your fat in the same places? Studies of families over many generations show that not only obesity is inherited but so is the type and location of the fat. One study showed the amount of abdominal fat was 56 percent heritable, which sounds like a pot belly gene.

Robert Hegele and colleagues at St. Michael's Hospital in Toronto compared waist-to-hip ratios to see who was apple-shaped (fat around the middle) and who was pear-shaped (fat carried lower). They studied the Hutterites, a small genetically isolated population in western Canada, all descendants of a utopian community established more than 100 years ago by a man named Joseph Hutter. The researchers found a genetic marker linked to waist-hip ratio in men but not in women. The gene appears to add several inches to male waistlines. Although it's not clear how the mechanism works, the marker is in a gene involved in the constriction of blood vessels.

The good news is that body weight can change, and people do successfully defy their genes. And despite the dominance of genes, environment always matters. No one, no matter what kind of genes they have, is going to get fat if they don't eat more calories than they burn.

The power of the environment is obvious in a place like mainland China, especially the poorer regions in the west. The Han Chinese, who live mainly on rice, are small and skinny. But when those same people move a few thousand miles to a richer society such as Hong Kong or Taiwan, they fill out nicely. So nicely that when they return to the motherland, they can barely fit onto the seats of trains built for the skinny natives. In our own hemisphere, there is a vast difference between the Pima Indians living in Arizona and their genetic kin in northern Mexico. The two groups were separated about 1,000 years ago, and the American Pimas have had every advantage in diet, health care, and ease of life. The difference in body weight is shocking: the Arizona Pimas have the world's highest reported prevalence of obesity, whereas the Mexican Pimas are just about average. The fatter Arizona Pimas also have significantly higher rates of diabetes, cholesterol, and high blood pressure.

This means that genes by themselves are not enough to tip the scales. If they were, all Han Chinese would be built the same way, and the Mexican Pimas would be as fat as the Arizona Pimas, regardless of diet. Nor is environment by itself responsible, because then everyone in Arizona would be fat and everyone in northwestern Mexico would be skinny. It's the combination that adds up. Apparently the fat genes carried by the Pima or the Chinese are only expressed in conditions of relative wealth and leisure. The bottom line is that genes are very important, but still dependent on the environment. A person who might balloon to obscene fatness in one setting could stay relatively thin in another. To understand how genes and behavior work together to control body weight, it's necessary to understand what the genes are doing.

SEEKING BALANCE

Body weight depends on a simple, unalterable balance: caloric intake versus energy expenditure, how much is eaten versus how much is burned. If you eat more than you need to live, you gain weight; eat fewer calories than you burn, and weight drops. Genes play a role in every part of the process: what makes us start eating, how we stop, and how calories are burned.

The body and the brain work together to regulate food intake. The key part of the brain is the hypothalamus, the same region that is so important for emotion, personality, and sexuality, so it's no wonder that eating is complicated emotionally. Clearly, there are many hungers being satisfied when we put things in our mouths. The main "start eating" center is contained in the lateral part of the hypothalamus. When this part of a rat's brain is removed or destroyed, the rat refuses to eat, and eventually starves to death. When it's stimulated by an electrical current, the animals eat even when they're full. This region is connected to the dopamine circuit, which is so important in rewarding behavior.

The brain's "stop eating" center is contained in a different part of the hypothalamus, the ventromedial region and paraventricular nucleus. When this part of an animal's brain is stimulated, feeding is inhibited. When it is destroyed, animals eat excessively until they are grossly obese. The same happens in humans. Alfred Frolich, a Viennese physician, first observed that cancer patients with tumors near the hypothalamus became obese. By using different brain chemicals, this region not only controls how much you eat but what you eat. For example, serotonin, which also is involved in worry and anger, inhibits eating carbohydrates; it's the satiety signal for spaghetti, but not meat balls (protein) or Alfredo sauce (fat). Norepinephrine stimulates carbohydrate intake. Finally galanin, a protein hormone, increases fat intake. The brain also controls the body's metabolism by signaling the nervous system. For example, the ventromedial hypothalamus, or stop-eating center, controls the vagus nerve, which stimulates the secretion of insulin, the master hormone of sugar and fat metabolism.

From the time we are born, hunger is a strong motivator of behavior. Babies scream for food, adults go to a restaurant, raid the refrigerator, or visit the vending machine. We usually experience hunger as a burning, growling "empty stomach," and being sated as having a "full stomach." This suggests that the physical sensation of an unfilled stomach sends a message to the brain to start eating, whereas the pressure of food against the wall of the stomach sends a stop eating message. To test this theory, a dedicated scientist named A. L. Washburn trained himself to swallow a balloon that could be inflated or deflated through an attached hose. He reported that when the balloon was inflated, he felt full whereas when the balloon was deflated he felt hungry. This supported the fullness theory. Washburn's heroic efforts were for naught, however, when others noticed that people who had their stomachs entirely removed because of cancer or severe ulcers, still felt hunger pangs. So there must be some additional signal—a chemical messenger rather than mere physical pressure. This so-called satiety signal has become the holy grail of obesity research.

One theory was that the brain was sensing the level of nutrients in the body, most likely the level of blood sugar, which rapidly rises with eating and falls with fasting. Another possibility was that the brain somehow knew how much fat the body was carrying. The "fat signal" was the object of much attention because it would be a perfect way to control weight and would make the discoverer extremely rich. So far, the best explanation of the satiety signal comes from studying fat mice.

OBESE MICE

As soon as you step onto the island that houses the Jackson Laboratories in Bar Harbor, Maine, one unmistakable smell hits you in the face: mice. The lab is home to the largest colony of mice in the world. There is not much diversity in lifestyles here; all mice are created equal and live pretty much the same lives, locked in cages eating compressed grain pellets and sipping water. Because they share the same environment, the mice are good indicators of genetic differences. If one strain of mice is unique, it must be in their genes. In 1950 geneticist George Snell identified a strain of mice that seemed to be pulled irresistibly toward obesity. They not only ate too much, they didn't exercise and couldn't burn calories. The obese mice also suffered from one of the most common problems associated with human obesity, diabetes. They spent their days sitting and fluffing their fur and were far less active than normal mice. They moved less, burned less oxygen, and had a low body temperature. In short, they were cage potatoes.

Despite the shared environment of all the mice on the island, these particular mice were so different that obviously something was wrong with their genes. The mutation was named "obese." It's a recessive mutation; both copies of the gene have to be bad to have an effect. To find out how the gene worked, another scientist at the lab, D. L. Coleman, sewed an obese gene mouse together with a normal mouse, so they shared the same blood. The result was that the obese mouse returned to normal weight. This meant there was something in the blood of the normal mouse, but missing in the obese mouse, that said "stop getting so fat." As a weight reduction plan for humans, stitching a fat person to a skinny person was not a terribly practical method, so the scientists set out to isolate the gene itself.

Jeffrey Friedman, a scientist at Rockefeller University, spent eight years mapping the obese mutation, isolating the small section of DNA and pinpointing the specific gene. He needed to find a gene that is expressed in fat cells, which is where an obese gene would logically be. He tested the likely candidates and found one that contained a "stop" signal in the middle, which meant that the gene turned out a shortened protein that didn't work properly. He found another mutation that also prevented the same gene from carrying out its designated task. He knew he had the right gene because no matter how its expression was disrupted, it produced obese mice. Somehow, a breakdown in this gene was making the mice fat. Further experiments showed the obese gene codes for a hormone, which was promptly named leptin, from the Greek word for thin. Leptin is an amino acid protein that contains a short sequence at one end that signals it to be secreted from fat cells and enter the bloodstream. The gene is expressed mostly in fat cells, and the fatter the cell, the more leptin it produces.

To prove that leptin really was controlling body weight, pure leptin was injected into obese mice. Sure enough, they started losing weight within four days, and by one month were 40 percent lighter. Further experiments showed that leptin had several different effects: first, the mice stopped eating so much; food intake was reduced by more than twofold. Second, the mice started burning their food more efficiently; energy metabolism was up, blood insulin and glucose levels were down. And third, the level of physical activity increased. So this one hormone could fix everything wrong with the obese mice: their eating habits, their metabolism, and their lack of activity. The leptin works this way: when stores of fat increase, the fat cells produce leptin, which tells the brain that it's time to stop eating and increase activity. When stores of fat decrease, leptin also goes down, which signals the brain to counteract the weight loss by increased eating and less activity.

THE HUMAN OBESE GENE

If the leptin mechanism could be manipulated in humans, it would be a breakthrough in weight control, which is why one company reportedly paid $20 million for the commercial rights to the obese mouse gene. When the human leptin gene was isolated, it appeared to be nearly identical to the mouse version. Then human leptin was produced artificially and injected into obese mice; it made them thinner.

A harder test was whether leptin would work on normal mice, not just on the genetically obese strain. When normal mice were injected with leptin, they did lose weight and eat less, but not as dramatically as the obese mice. They also needed large doses of the hormone, which suggested that normal people would need to be injected with large quantities of leptin to have any effect on weight. This meant leptin wasn't going to be a very good artificial stimulant to weight loss, but there was one other hope: what if people were obese because their leptin gene was faulty? Perhaps fixing the gene could help them reduce.

To test that idea, fat cells and DNA from more than 200 obese people were scanned. The initial results were disappointing: the obese subjects were actually producing more leptin than normal and didn't appear to have any mutations in the gene. Sadaf Farooqi, a researcher at Cambridge University, decided to narrow the focus and look at the DNA from the two most obese subjects he'd ever seen: a two-year-old boy who weighed 64 pounds, and his cousin, a 190-pound eight-year-old girl. Sure enough both cousins had the exact same mutation in both copies of their leptin genes. It was a mutation that disabled the leptin gene, producing an inactive version of the protein. This was the clear proof that leptin was a key player in human weight control, just as in mice.

The most striking thing about the leptin gene mutation in the cousins was how rare it was. Clearly, obesity in most people, even those like Sandra, was not caused by a mutation in the leptin gene. But there was another possibility: perhaps being overweight boosted Sandra's leptin level. This theory was tested in Finland on 23 pairs of twins, one fat and one skinny. The obese twins had leptin levels three times higher than their skinnier twin brothers and sisters. In another experiment, leptin levels were measured in people before and after they had lost weight by dieting. As weight dropped, the leptin levels fell, in some cases by more than fivefold. Obviously the leptin gene hadn't changed because it lacked fudge cake, so the decrease must have been physiological, not genetic.

Scientists weren't ready to give up on a genetic key to obesity, partly because they knew about other mutations that caused similar weight problems in rodents. For example, a mouse mutation called "diabetic" causes obesity, frequent occurrence of diabetes, poor metabolism, and lack of activity. A crucial difference between the diabetic mice and obese mice is that the diabetic mice don't respond to outside leptin. When a diabetic mouse is sewn together with a normal mouse, the normal mouse shuns food and eventually starves to death. The interpretation is that the diabetes gene makes the leptin receptor: a brain protein that senses the presence of the fat hormone. If the receptor detects leptin, it tells the body to stop eating and start burning fat.

When the leptin receptor gene was isolated, first from mice and then from humans, it was found to be located primarily in the hypothalamus, the brain's eating center. When fat cells are well fed and plump, they produce leptin which goes to the brain and binds to the receptor, which in turn decreases food intake and increases metabolism. Thus, the leptin receptor must send signals both within the brain to decrease hunger and to the body to increase metabolism.

METABOLISM

The leptin receptor signals the body to increase metabolism through a circuit that leads from the hypothalamus to the spinal cord, and from there to the nerves that stimulate fatty tissue. The nerve terminals surrounding the fat cells release norepinephrine, which interacts with something called the β3-adrenergic receptor.

Initial evidence that the β3-adrenergic receptor might be involved in obesity and diabetes came from two observations. First, activating the receptor with norepinephrine-like drugs causes fat cells to burn off fat. Second, the fat cells of obese mice have 300 times fewer β3-adrenergic receptors than the fat cells of normal mice, and the few remaining receptors don't function normally. To test the theory in humans, an international team looked at the largely overweight Arizona Pima Indians. Could this receptor be faulty in Pimas and causing them to be so overweight? The scientists chose ten Pimas suffering from both obesity and diabetes and screened the DNA for the β3-adrenergic receptor gene. The researchers were astounded to find that five out of the ten had the same mutation.

Studying a larger group of 642 Pimas showed the mutation did not strictly determine whether an individual got diabetes but when. People with two copies of the mutation (one from each parent) got diabetes at age 36, which was four years earlier than people with one copy of the mutation, and five years earlier than people without the mutation. Furthermore, people with the mutation tended to have a lower resting metabolic rate than those without the mutation; they were burning less energy, and their fat cells were not responding to the signal sent from the brain. The same mutation has been found elsewhere with the same effects. In a study in Finland, for example, people with the mutation had an earlier age of onset of diabetes than those without the mutation.

The Finnish study also looked at the effect of the receptor gene on body shape. When brothers and sisters were studied, the one with the mutation had a higher waist-to-hip ratio than the sibling without the mutation, even though they had grown up in the same home and at the same table. A likely explanation is that this particular receptor is most active in the brown fat cells that surround the gut.

The receptor's impact on the waistline may also be partly responsible for middle-age bulge. Some weight gain occurs with age in most people, but a study of middle-aged people in Paris also pointed to the receptor. The study compared 185 morbidly obese people to 94 people of normal weight. Over a 25-year period, from age 20 to age 45, people with the mutation gained 30 pounds, while those with the more common form of the gene gained 23 pounds.

Sandra would have been a good candidate for the β3-adrenergic receptor gene mutation. Her profile is typical of people with the mutation, and it could explain why her obesity and diabetes hit so quickly. What if she had known about the mutation? What if the doctor had scanned her DNA and notified her of the problem? Judging from Sandra's behavior, probably nothing would have happened differently. She already had a large genetic warning in front of her—her mother—but she didn't eat less or exercise more. Perhaps the specific information about a gene would have made Sandra sit up and take notice, just as a first heart attack focuses many people on strict adherence to a diet. Knowledge without behavioral change is worthless, however, because at this point there is no genetic fix for the mutation, no way to go in and change the receptor. The only solution is to use the information to change behavior. Knowledge of nature must be used to change nurture.

FEELING HUNGRY

Like it does with metabolism, leptin works on hunger indirectly through a second messenger. In the case of hunger, the messenger is a molecule that's produced in response to leptin called neuropeptide Y, which acts as an appetite stimulator. When this substance, known as NPY, is infused into the lateral hypothalamus, the brain's start-eating center, it causes ravenous eating. Rats injected with NPY will eat food even when it's been laced with quinine to make it taste horrible. Under the influence of NPY, rats will lap up milk when they know it will mean an electrical shock to the tongue.

Richard Palmiter and colleagues at the University of Washington bred a strain of mice without the NPY gene. These mice should have been skinny, since they lacked the chemical that told them to start eating, but they appeared normal. However, when these mice were crossed with the obese mice (who have no leptin), the lack of NPY had a big effect. The mice were still plump, but only half as fat as the regular obese mice. The experiment showed that NPY is part of the reason that leptin-free mice eat so much, but not the whole reason.

The bad news of these experiments is that they establish an invariable link between how much we eat and how well we burn calories. Metabolism and hunger are branches from the same pathway that begins with leptin. This is why the people who are born to overeat also burn calories slowly; the two traits go together because they come from the same source. This makes sense from an evolutionary perspective because these people would have an advantage during famine, but in the current American abundance they are the ones who must fight hardest against getting fat.

Still other genes control what food we enjoy, and some people are irresistibly drawn to foods that are not always good for them. This was discovered by accident in 1931. Dr. Arthur Fox, a chemist at Dupont, had just synthesized a new compound called phenylthiocarbamide, or PTC. There was an explosion in the lab that scattered chemicals everywhere, and one scientist accidentally got some PTC in his mouth. The compound was harmless, but the scientist complained of the bitter taste. When Fox himself tried it, he didn't taste a thing. Excited as only a real scientist could be, Fox decided to do an experiment. He passed out samples of PTC at a meeting and asked everyone to taste it. Twenty-five percent said they couldn't taste anything, half found it somewhat bitter, and the remaining 25 percent found it unpalatably bitter. Fox concluded that he was among the small number of nontasters, and that people with normal taste buds would find the substance bitter. The small group who couldn't stand the taste were dubbed supertasters.

When entire families were studied, the ability to taste PTC was traced to a single gene. Later research showed that the gene regulates the number of taste buds, which send nerve signals to the brain centers for taste, pain, temperature, and touch. The differences were dramatic: about 1,100 taste buds per square centimeter in supertasters compared with just 11 per square centimeter in nontasters. This simple genetic difference in taste buds has a profound effect on what tastes good and what tastes bad. Supertasters find very sugary food such as frosting too sweet, fatty foods such as cream too greasy, and many healthy vegetables such as broccoli and mustard greens too bitter. Supertasters prefer their food tepid rather than steaming hot or icy cold. Nontasters can eat anything and everything, sweet or greasy, plus hot peppers and sharp spices such as ginger. Supertasters, who are turned off by sugary and fatty foods, tend to be thinner and have lower cholesterol among older women, according to research by Drs. Linda Bartoshuk and Laurie Lucchina at Yale University. These thinner women aren't especially virtuous or strong willed; they just don't enjoy eating what's not good for them.

Fat's Not All Bad

Most people do enjoy sweets and gooey treats, which is why we have donut shops and ice cream parlors instead of broccoli stands or grapefruit bars. Eating rich or sugary foods wouldn't be such a problem if the body didn't turn them into fat. In some parts of the world, fat is still what gets people through the winter or hard times, but for modern Americans fat is a menace. Now that we live in such abundance, why do we need body fat at all? Instead of storing fat in a big roll around the waist, why can't we store it in the refrigerator in the form of food? Since we have food when we need it, we theoretically shouldn't need to wear it under the skin.

Chuck Vinson and his colleagues at the National Cancer Institute developed a strain of mice almost completely without fat cells. Predictably, the mice were skinny, so skinny that they kept dying off, almost as if they were shivering to death. That problem was solved by lining their cages with heating pads. There was another downside to the skinny mice: their personalities. They were terribly jumpy and mean. They didn't run around much on their own, but when someone tried to pick them up, they hopped and squirmed. It was easy to know which mice were the fat-free variety because they were the hardest to catch. They also attacked without provocation, and the cage was often splashed with blood and fur. Vinson, rubbing what he calls his own "Buddha belly," says with a big smile, "Fat and happy, lean and mean."

Fat and happy certainly is the stereotype, and the bigger the better when it comes to jolly Santas or dancing bears. But there are too many unhappy fat people, and too many cheerful thin ones, to positively link personality types with body sizes. There are, however, certain traits that can predispose a person to overeat, especially in certain situations. Since our genetically influenced personalities have so much sway over our behaviors, and because eating is such an important behavior, it would be surprising if personality didn't have some link to body weight. Some genes influence weight directly by controlling metabolism or hunger, but others work indirectly by shaping personality traits that in turn influence how, what, and how often we eat.

Emotional Overeating

In the case of Sandra, it is impossible to say if she had a biological problem that made her fat. One thing that is certain, however, is that she used food as an emotional support, and that as the stress in her life increased, so did the amount of food she consumed. She didn't physically need more calories, but she was feeding an insatiable emotional hunger.

Consider the case of George, a successful college professor, proud husband and father of five, and devout Mormon. George had worked hard all his life, and everything he had, he had earned. His career was exemplary, and he was rewarded with the department chair at the young age of 40. He was equally determined to be a good husband and father, which meant that he seemed to be getting up earlier and staying up later as his family grew. Soon he was rising at 5 A.M. in order to make the long drive to the university and be at his desk at 6:30, where he would work furiously until 4 P.M. He always left promptly, even if he had to stop writing in mid-sentence, to get home for dinner with his family. After dinner, homework, chores, and family activities, his wife would put the kids to bed, and George would open his bulging briefcase to continue working into the quiet of the night. His wife would come down again, serve him a little slice of pie, and kiss him goodnight.

George had one weakness: snacks. George ate breakfast early each morning, which meant an early lunch too. By the time he was ready to drive home he was famished, so he'd grab a snack for the ride. He'd turn on the radio and crack open a box of cookies, or a bag of donuts, or a couple of candy bars. He found the snacks helped him unwind. By the time he pulled up to the house, he had forgotten about work and could focus on his family. Life at home was wonderful, and he loved the scramble of activity as the children clamored to greet him before he could struggle out of his coat. He'd usually have one on his shoulders and one in each arm as he walked to the table, which was piled high with healthy and hearty food. George said grace and the family began to eat and talk of the day's adventures. Life truly was good.

When George took over the department, he was about 10 pounds overweight, not fat but not as thin as when he was in college. He used to jog in the mornings, but there was no way to do that and still get to work early. Then he tried working out at lunch, but a student or a colleague invariably would interrupt his plans. Almost without realizing it, he had become sedentary. Two years later, he had to buy a new, bigger wardrobe. He took to wearing suspenders instead of belts, and he left his shrinking tweed jackets hanging in the closet. In a few more years, George was obese. He wasn't unhappy, and he wasn't terribly self-conscious about his weight. He thought of himself as big rather than fat, and he didn't believe people should be judged by their shape. George knew that he was eating because of the stress, but he figured that was a lot better than smoking or drinking.

People like George are fascinating for Judith Wurtman, the head of the Nutrition and Behavior Studies Group at MIT's Clinical Research Center. She calls them "emotional overeaters." Wurtman believes there are two types of hunger: physical hunger, which is a response to the body's need for nourishment, and psychological hunger, which is "propelled by our need for comfort and solace." The key to psychological hunger, according to Wurtman, is serotonin—the same neurotransmitter involved in the personality traits of neuroticism and harm avoidance, and aggression and hostility. Serotonin is the key chemical involved in feeling depressed, anxious, tense, irritable, frustrated, angry, stressed, and mentally exhausted. Physical hunger can be satisfied by eating a wide range of nutritious foods; psychological hunger can be satisfied only with foods that supply serotonin.

For people with these kinds of cravings, it would seem logical to eat high-serotonin foods, or better yet a no-calorie serotonin pill. The problem is that serotonin can't pass directly from the bloodstream into the brain. Instead, the body needs sweet and starchy carbohydrates to boost the brain's supply of tryptophan, the amino acid building block for serotonin. The system is even more complicated because carbohydrates don't actually contain any tryptophan. When carbohydrates are digested, insulin is produced, which pushes glucose and all the amino acids except tryptophan into the cells to be used for energy. When everything else is gone, the relative concentration of tryptophan left in the blood is higher, so when the blood enters the brain it is turned into serotonin.

Paradoxically, proteins have just the opposite effect on serotonin. Unlike carbohydrates, proteins do contain tryptophan, but it's the least abundant amino acid. That means eating a sirloin fills the blood with so many other amino acids that the tryptophan is kept from entering the brain. It's like trying to get on a train that's already filled; there simply is no room on board. There's always room on the carbohydrate train, which is why when you feel down or depressed, the first food that comes to mind is a cookie, piece of cake, or potato chips. If you're watching your diet, you'll feel like a bowl of pasta or a bagel before a pork chop or beef jerky. The brain wants its serotonin, and the best way to get it is to load up on carbohydrates. Another way is through drugs, which explains why dexfen-fluramine, the active ingredient in the diet drugs fen-phen and Redux, targets serotonin.

At MIT's Clinical Research Center, a typical group of subjects ate normally and moderately at mealtime, but consumed up to 1,000 calories in snacks, most often two to three hours after lunch or dinner, and they always chose carbohydrates such as crackers, candy bars, chips, and cookies, and not proteins such as yogurt, cheese, cold cuts, or even hot dogs. The subjects said they usually snacked when they felt distressed, down, tired, or anxious. Their eating pattern would support the serotonin explanation, but it could also be that they just prefer the taste of cookies to cold cuts or have some past psychological association with certain foods that makes them comforting.

As an experiment, they were offered a drink rich in carbohydrates and given a series of questionnaires and interviews to measure mood. Later they were given an identically flavored drink, but this one rich in protein, and the same mood tests. The protein drink did nothing for them emotionally, but the carbohydrate drink made them feel less blue and more peppy. Further support for the serotonin connection came when some were given dexfenfluramine and others were given placebos. Those taking the serotonin drug cut their snacking in half.

Our own studies at NIH showed a link between personality traits associated with serotonin and being overweight. We collected height and weight information and personality profiles on more than 500 people. The subjects were being studied for things other than their weight, which varied from skinny to obese. When we compared relative body weights to personality types, the strongest correlation was for impulsiveness, which is part of neuroticism or emotionality. People who scored high perceive desires to be too strong to resist, even if they later regret their actions, while low scorers find it easier to resist temptation. The body-mass index also correlated, but more weakly, with depression and angry hostility. The result suggests that emotional sensitivity is playing a role in eating. Or perhaps the reverse is true, that eating plays a role in how we respond to our emotions.

What these experiments show is that there is a connection between eating habits and certain traits, such as neuroticism, harm avoidance, stress, anger, and depression. The traits tend to be the ones mediated by serotonin, which means that eating habits also are partly mediated by serotonin. The genes that control this have not been found yet, but the search is on.

EATING DISORDERS

Thirty years ago at the Jackson Laboratories, famous for its mice, there was a college student in the summer research program whom we'll call Beth. She was so thin that her ribs showed beneath her skin, and her arms and legs were pale sticks. She dressed in T-shirts and shorts, which flapped around her delicate limbs. She worked hard and was a good enough student, not really any different from the rest of us, except for one thing: Beth never ate much at meals; everything seemed to disagree with her. It was rumored that she had some sort of disease—all of us junior scientists assumed it was something metabolic—that made eating difficult. The really strange thing was that she always took food with her from the dining hall. She wouldn't eat anything at lunch, but she'd make a big ham sandwich, wrap it up, and put it in her pocketbook to take to the lab. Or at dinner she would meticulously wrap a piece of chicken and take it. Her lab refrigerator and dorm room were full of food packets, all carefully wrapped and hidden away.

With hindsight it seems obvious that the disease Beth had is anorexia nervosa. This condition is characterized by an intense fear of eating, often to the point of becoming dangerously, sometimes even fatally, underweight—essentially starving to death in an effort to be ultrathin. The disease is seen almost exclusively in women, especially adolescent girls. As many as 1 out of 100 teenage girls may have anorexia. Over the years many different causes for anorexia nervosa have been claimed: unrealistic expectations of female body shape fostered by the media and popular culture, childhood abuse, sexual repression, etc. In fact anorexia is almost certainly a biological disease that represents a distortion of a natural biological response—the response to famine.

The tip-off is the hoarding behavior. Imagine that there was a famine and no one was able to obtain food. The natural response, undoubtedly selected by evolution, would be to squirrel away any food that could be found. In anorexia, the trigger is not famine but a deliberate, voluntary effort to lose weight; anorexia almost always starts with a diet in which the girl decreases food intake to lose a few pounds. But when she loses the weight and finds her life is not changed, she just keeps on going, trying to lose more and more, while hoarding food according to her natural instinct, responding just the way her ancestors did to famine. The distortion is that she doesn't eat the available food.

Thus, according to this model, people with anorexia are simply taking to extremes a bodily response to food deprivation. In support of this, a study by J. Hebebrand at the University of Marburg in Germany has shown that women with anorexia have lowered leptin levels, as would be expected for someone actually starving to death. The question is not why the behavior starts, but why it doesn't stop. Here personality probably plays a key role. Twin studies show that anorexia has a substantial genetic component, and moreover it is part of a spectrum that includes neuroticism, depression, low self-esteem, and general affective instability. So, in looking for genes for this puzzling and all-too-common disorder, both genes related to eating, such as those involving leptin, and those involved in emotionality, such as serotonin, would be obvious places to start.

The other major eating disorder is bulimia nervosa, which is characterized by a loss of control of eating. The name comes from the Greek words for ox and hunger, and the typical behavior is to gorge, especially on snacks or dessert, then to purge with vomiting or laxatives. Interestingly, the gorging most often occurs in the late afternoon or evening—exactly the same time when some obese people crave carbohydrates. This suggests that bulimia may be an extreme form of the type of emotional overeating studied by Wurtman at MIT, which has been related to anxiety and the serotonin system. Indeed, serotonin drugs have shown some success against this disorder.

BEING THIN IN A FAT SOCIETY

Paradoxically, we live in a society that worships thinness but continues to grow fatter. Popular images abound with "beautiful" people who boast rippling abdominal muscles,

and wispy fashion models who appear frail or even malnour-
ished. Yet if you actually sit down at an airport or shopping
mall and watch real Americans pass by, the surprising thing
is how fat they are. The national weight is climbing, perhaps
dangerously so. We are carrying the fat of the land on our
swelling bodies. Much of the problem is the result of our
success as a species. We have conquered nature and divided
our labors so that some people never have to lift a finger to
survive. Only a few thousand years ago, most humans ex-
pended great physical energy just to provide food, clothing,
and shelter. Today we have divorced food from hunger and
activity from survival. Now we eat for many reasons, not just
to stay alive, and we have created special places, designed for
maximum comfort, where we engage in physical exercise. In
a society where much work is performed sitting down, "leisure" means working out.

Our obsession with being thin actually contributes to
making some people fatter. The reason is that a drastic or
crash diet causes the body to fear starvation. The natural,
physiological response to a sudden shortage of food is for the
fat cells to decrease the production of leptin. This sends a
powerful bodily signal to eat more and to conserve calories,
which is why yo-yo dieting not only fails to keep weight off,
it can result in weight gain.

Here is a case where society's success brings new and
unexpected challenges to individuals. It's great that we can
run out for hamburgers, instead of running down a wild
boar. It's nice to watch television, instead of chopping wood
or fetching water. But the risks are enormous to people genetically built for the Stone Age. Even people with "normal"
genes must struggle with the confusing mix of signals bom-
barding us with new taste treats, while idolizing lean bodies.

Perhaps a little moderation is in order. People do need to watch what they eat and to exercise. We do need to prevent obesity, but we needn't agonize over normal ranges of body weight or appearance. It would be nice to ease up on the social pressure to be thin, especially for adolescents who are the most vulnerable emotionally. We have invested too much of who we are with the size and shape of our bodies. Wouldn't it be wonderful to see late-night TV commercials that didn't advertise machines to ripple your abs, but ones that exercised what's between your ears?

EIGHT

AGING

The Biological Clock

Those wishing long lives should advertise
for a couple of parents, both belonging
to long-lived families.

—OLIVER WENDELL HOLMES,
JUSTICE OF THE U.S. SUPREME COURT

Martin and Beatrice were married during the early years of World War II, and like many young couples of their generation, they soon were forced to say good-bye. Martin proudly and eagerly put on an Army uniform and was sent to Europe to fight. Beatrice would not have had it any other way.

From the time when he was young, Martin was a precise, orderly person. Although he was from a modest home, he was always impeccably dressed and groomed. He loved wearing his uniform, with the sharpest possible creases, and he enjoyed the discipline of the Army. In the Army, he found the perfect outlet for his sharp mind, keen memory, and desire that everything be just so. He was assigned to be a topographical engineer, one of the men sent in before an attack to map beachheads and invasion routes. Thousands of lives depended on Martin's accuracy and precision. He was good at his job, and after serving in the European theater, he was sent to the Pacific to finish the war.

After the war, Martin continued his work with maps, but as a civilian at a senior level in the Pentagon, and he served with distinction for 35 years. He and Beatrice had scrimped and saved to afford an almost new home on a quiet street in northwest Washington, D.C. Two bright and healthy sons were raised with the firm hand of Martin and the loving embrace of Beatrice. Their yard was meticulously kept, and an American flag hung off the porch on holidays. Beatrice managed the house, and when the boys were old enough, she took a job at an insurance company, where she was a valued and loved employee for nearly 28 years. Every year, they managed to set aside enough money for a family vacation.

Martin was only 55 when he retired. He kept busy with volunteer work, scouting, and the church. He was still a powerfully built man in excellent physical condition, except for the onset of diabetes late in life, which he controlled through diet and insulin shots. Martin had no trouble at all following the strict diet. He read up on diabetes and ate only what he was supposed to eat. There was no fudging, no secret snacking. Martin pursued his diet with the same discipline and attention to detail with which he pursued every course of action. He credited his good living to the fact that he felt hardly any different from when he had shipped off to Europe. He had mellowed and matured, of course, and he didn't fly off the handle as much as when he was younger, but time certainly had not dulled the creases in Martin's trousers.

Looking back on their last years together, Beatrice says

the signs should have been unmistakable. Their sons noticed something wrong with Martin, but she refused to see it. Sure, there were little things. She heard him once outside at 2 A.M., fidgeting with the keys and cursing at the car. When he finally came inside, and Beatrice questioned what he was doing, he exploded. "I felt like moving the car, that's all," he said. Then there was the money. He had to have $50 bills on him at all times, and every week he was withdrawing several hundred dollars from the bank. He always had been generous to a fault, but now he was passing out money to friends and relatives and anybody else he came across. Beatrice had no idea how much he gave away, or to whom. A family friend who worked at the bank asked Martin what he was doing with all money, and he whispered conspiratorially: "My wife is stealing from me."

The gentle arc of aging that Martin's life had been following began to point downward, steeply. The fall was abrupt, and soon obvious even to Beatrice, who had refused to accept that her big, strong, six-foot man, a self-made man who had earned everything he had, was stumbling. As a result of his forgetfulness, he became lax in controlling the diabetes, and his health began to suffer. Once he sat in a chair and spooned out big bites of ice cream, something he hadn't eaten in decades. After gorging, he got up and put away the carton, only to open it again 20 minutes later. A half hour later, he got up for more ice cream. After the third or fourth time, Beatrice finally said something. He shouted at her: "What are you doing begrudging me a little bit of ice cream?" and hurled the carton across the room.

Beatrice knew that their lives were heading into a new phase. Their "golden years" were not going to play out exactly as she had planned. They had hoped to travel and visit their grandchildren, remain active in the church and the community. Now she began to take stock, emotionally and financially, and to mentally prepare herself. She had no earthly idea how bad things were about to get.

We spend our lives thinking about aging, but how we regard the process depends upon our place along the time line. Children count the weeks until their birthdays, and young people can't wait until they are old enough to drive, or drink, or stay out late. In our twenties we are invincible and oblivious to passing years, until the thirtieth birthday. Women, especially, hear the loud ticking of biological clocks as their reproductive years come to an end. Middle age sneaks up on us, as we continue to push ahead our idea of what constitutes "old." We compare our achievements to those of other people until one day we realize that there are a lot of people who are younger and more successful. Priorities change, our bodies make more demands, and at some point memories of the past outnumber plans for the future. Suddenly we are old. And then we die.

People think of the human body as a machine, like a car that can go only so many miles and hit so many potholes before it wears out from use. The tires go bald, the valves gum up, the battery loses its charge, and the shock absorbers lose their spring. Most parts can be replaced as they wear out or break down, but there comes a time in the life of a car, and a person, when piecemeal repairs are not enough and the whole thing simply shuts down.

The recent discovery of several key genes related to aging suggests that the car analogy is a good one, but not for the obvious reason. The body doesn't wear out from use as much as it is designed to last only a short time. The reason car companies stay in business is that cars need to be replaced every few years; they are built with planned obsolescence. The human body is built the same way; our parts wear out not simply from use but by design. We die from planned obsolescence. Our genetic blueprint comes with fine print that reads: warranty valid only for limited time.

The notion that we are programmed to die seems odd. If evolution is so efficient, then why don't genes select for longevity? In other words, people who are healthier and live longer should pass on their genes, continually extending the life span of the species. We have in fact extended our life span, but not because of genes. The additional years have come quite recently and entirely from our own deliberate improvements in our environment, such as better diets and health care. There is no evolutionary reason why genes should have changed to make us live longer. There will always be a certain fraction of any population that dies because of purely environmental causes, such as accidents, predators, or disease, so the genes that really matter are the ones that favor survival early in life. Once the organism has survived past the age of reproduction, it is useless as far as evolution is concerned. The genes don't care if you live one day after you deposit your sperm or egg in the bank of future generations.

We aren't really programmed to die; we are programmed to live only long enough to reproduce. Death may provide some benefit to the species, but the reasons are pure speculation. Perhaps the benefit of death is that it frees up resources for the young and fertile and keeps churning the gene pool. Or maybe it is the specter of death that encourages us to reproduce and thus live on through our children. But whether or not death has a purpose is mostly irrelevant to the continuation of the species. Genes that make us strong and healthy during our reproductive years are always more advantageous to the species than are those that might make us live longer.

As individuals, there is no doubt we want to live longer, or at least look younger. We have a strong instinct to survive combined with an acute sense of mortality, which has pushed us to improve living conditions that prolong life. Not to mention the power of human vanity. We still fantasize about discovering a "fountain of youth," but now the search for longevity has turned inward to look at our genetic code. By discovering the genes that control aging, perhaps we will be able to prevent or at least slow the process. More likely, we will be able to prevent some of the diseases, disability, and discomfort that accompany aging. Perhaps the most important result of discovering the genetics of aging will be to understand what cannot be changed. There are some aspects of mental and physical decline that cannot be avoided, but our enjoyment of our final years certainly can be increased through medical advances, healthy living, and positive thinking.

LONGEVITY GENETICS

The common folk wisdom is that people are likely to live long lives if their parents did. This popular theory that longevity is partially genetic has been proved correct by many scientific experiments. In a pioneering study at the Johns Hopkins School of Medicine in 1934, two scientists named R. Pearl and R. W. Pearl indentified a group of people who lived to be 90 or more and showed that a high proportion of their ancestors were likewise long-lived. Forty years later, another generation of scientists at Hopkins followed up on the children of these nonagenarians and showed that they too had on average a lengthened life span.

The best estimate of how strong a genetic role there is in longevity comes from a twin study. In a group of 2,872 Danish twin pairs born between 1870 and 1900, the age at death was closer for identical twins than for fraternal twins, showing a genetic effect. But it was not a terribly strong effect: the heritability of longevity was estimated to be 26 percent in males and 23 percent in females. Most of the variance in how long people lived was explained by those seemingly random factors unique to the individual and not shared within the family. Even that moderate heritability of longevity is probably spread out among many, many different genes. By one estimate, 70 percent of the 100,000 or so human genes have an influence on life span. That would make 70,000 "aging genes," far too many to study conveniently, especially since there has been so little raw data—only a few generations of people—generated since the science of genetics began.

Not everyone has been put off by the difficulty of finding genes for aging. In 1976, when Michael R. Rose was a graduate student in genetics at the University of Sussex in England, he started his "family" in a few milk bottles. He filled the glass bottoms with nutrients and inhabited them with 200 fertilized female fruit flies. After five weeks, the flies were at the end of their reproductive "years." Rose collected the eggs from those few who were still healthy and fertile, and bred a new generation. He waited five more weeks and again took the eggs from the oldest breeding flies. He kept repeating the process, each time selecting the progeny only of the longest-lived flies. Just as he'd hoped, at each new generation the selected flies lived a tiny bit longer than their ancestors.

Today, Rose is the proud father of a million flies, and they are still breeding, each time being selected for longevity. Rose now has 50 research assistants to help him with the project at the University of Irvine in California. The amazing thing is that the current generations of flies are living twice as long as the founding flies, and they keep living longer. If these were humans they'd be living to 140 years old. Rose calls his 120-day wonders "Methuselah flies."

Rose's experiments show two important things. First, that genes play a critical role in aging. His long-lived flies aren't doubling their life spans because of improved medications, health care, or seat belts. Only their genes have changed, and only "naturally" in that Rose did it with breeding rather than intervention. Second, there must be a very large number of aging genes. If there were just a few genes, Rose's experiments would have come to a halt a long time ago because the best genes would have been selected for after a few generations. The fact that Rose is still breeding older and older flies means there must be a vast, still not-fully-tapped reservoir of genes for aging. To tap that reservoir, scientists must find out how the aging genes work.

BODY RUST

Dr. Doug Wallace of Emory University School of Medicine shows some sobering slides about how we age at the cellular level. At 24 years old, the slides show a taut, well-formed band of color. At 33, the band is still distinct but the edges are smeared. As the years increase, the image blurs and the band appears to sag and degenerate. In the final slide, at 94 years old, the original band has disappeared completely, and nothing is left but a hazy, fuzzy, mess. The slides show how the years chew up the DNA contained in the mitochondria, the powerhouses of human cells. We start out life with a mitochondrial genome of precisely 16,569 bases, but as we age and our cells divide, there are mistakes, skips, and deletions. Since DNA can only be lost and not replaced, by the time a person is past 80 years old there is hardly a single genome still intact.

The loss of DNA is a sign of a more general degenerative process, oxidation. Every cell in the human body uses about one trillion molecules of oxygen per day, but not all oxygen is good. Forms of oxygen that contain an unpaired electron, known as free radicals, are among the most volatile and destructive toxins generated by the body. Because of their extra negative charge, free radicals promiscuously attach themselves to many different types of molecules. They attack DNA, proteins, and lipids, causing age spots on the skin and damaging cellular repair and reproduction.

When oxygen combines with metal it's called rust. When the free radicals generated from oxygen attack our cells, it's called aging. Our very breathing, which we need to live, is at the same time killing us by effectively rusting our bodies. When we are young and healthy our cells are able to repair and replace the damage caused by the free radicals, but as the oxygen attacks the mechanisms for repair and replacement, we become increasingly susceptible to the damage to cell metabolism and the damage to DNA that produces mutations including those that cause cancer.

The oxidation theory of aging makes a potent prediction: that genes that increase free radical production and destructiveness should accelerate aging, whereas genes that inactivate free radicals or improve the ability of cells to protect themselves should prolong life. So far that prediction has been best tested in an unlikely organism—a tiny worm.

This bacteria-eating roundworm, just one millimeter long, is a hermaphrodite, meaning it fertilizes itself, and lives for just nine days. The worm shares at least 40 percent of its genes with people, and its brief life is a mirror image, tightly compressed, of the human life span. As the worm ages, it first loses fertility. Then movement gradually slows down, and the ability to protect and repair against oxidative damage deteriorates. Mistakes and alterations accumulate in the DNA, especially the mitochondrial DNA, and death follows. Careful breeding produced a strain of super worms that lived five times longer than normal, the equivalent of a human living some 350 years. The feat was achieved by combining several of the different genes involved in aging.

The first life-extending mutation was found in a gene dubbed "age-1." Mutants with this gene live twice as long as normal worms, and they are just as healthy, mobile, and fertile as any other worm when they are young. As they age, the mutant worms show a much slower rate of loss of mobility and also of accumulation of mitochondrial DNA mutations.

To test the idea that the worms were living longer because the age-1 gene protected them from free radicals, the worms were exposed to high concentrations of oxygen or a chemical that generates free radicals. The mutants did prove tougher than normal worms on several counts. They were much more resistant to the free radicals, to heat, and to ultraviolet light, which may generate free radicals. The mutant cells make increased levels of two enzymes, catalase and superoxide dismutase, that can convert toxic free radicals into more benign molecules. So worms bred to have the age-1 gene were indeed more protected against oxidation.

A separate mutation slowed down the worms' biological clock. Worms with this gene, called "clock," develop more slowly than normal worms and live their lives at a slower pace. They do everything at reduced speed: embryonic development is delayed, cell division takes longer, movement and swimming are at a leisurely pace, even defecation is slowed down. And so is death. The life span is increased by a half or more. It's as if the worms were living in a world in which time itself had slowed down. One plausible explanation for the increased life span is that the gene decreases the accumulation of free radicals or increases the accumulation of enzymes that break down the toxins.

Even normal worms have more control over the pace of life than humans. When faced with a drought or famine, the worms can hibernate by transforming themselves into a dormant state. When conditions improve, they can return to their normal lives. Certain genetic mutations increase this hibernating state and double the life span. During the dormant period, the worms don't eat or move (or accumulate damage from free radicals), but they do stay alive. The record-breaking worms, the ones that lived five times longer than normal, were produced by crossing the long-hibernating worms with the ones with the slow clock.

These life-prolonging genes found in worms are important because they are connected with oxidative stress, the same mechanism thought to be involved in human aging. Whether the same type of genes will be found in humans remains to be seen, but the search is on.

Human Age Genes

A few key genes have been found in humans that relate to aging, and in one case, a single gene has been discovered that has a devastating effect on the life span. In 1904, a German physician named Otto Werner reported on a family with premature aging. People with this strange condition seemed to have a rapidly accelerated biological clock, so they lived an entire lifetime in a few years. Birth, childhood, adolescence, adulthood, and old age came in the proper order but were blurred together in a few short years. Werner speculated a genetic cause, and he was right. The condition, now known as Werner's syndrome, is caused by a mutation in a single recessive gene. The disease occurs only when a child inherits a mutated copy of the gene from both parents; individuals with just one "bad" copy of the gene show no symptoms of premature aging. But having two copies that aren't functional leads to the many symptoms of premature aging, from graying of the hair, to wrinkling of the skin, to cancer.

Genetic mapping placed the Werner's syndrome gene on the short arm of chromosome 8, but scientists were unable to pinpoint the exact location. An international team working on the search, led by Gerard Schellenberg of the Veterans Affairs Health Care System in Puget Sound, was stuck. The difficulty was that they didn't know what the gene did, so they didn't know what kind of gene they were looking for. The only solution was to determine the precise sequence of every single gene in the suspected area, a total of 1.2 million bases. The painstaking search paid off when they found a tiny segment of DNA that was different only in people with the syndrome.

Even when they had located the gene, they still didn't

know what it did. Using a powerful computer, they compared the mystery gene's DNA sequence to those of all known protein-coding genes in the entire DNA database—a collection of some 1.5 million sequences of every type of organism from the simplest single-cell bacteria up to man. This was like running a set of fingerprints through every law enforcement database in the world, only this database includes other species as well as humans. The DNA code came up positive for a gene also found in worms, yeast, and in a kind of bacteria.

The gene is a helicase, an enzyme that unwinds the double-helical DNA molecule. Think of unwinding a circular staircase to make a ladder. The helicase opens up the DNA molecule to expose it to enzymes that catalyze various reactions. The staircase can't unwind in people with Werner's syndrome, and DNA metabolism is a disaster: whole chromosomes are lost as the cells divide, there are strange crossovers of DNA sequences and increased mutations. The housecleaning enzymes can't get in to repair the normal wear and tear, so the damage accumulates. The most likely reason that Werner's patients so frequently get cancer is that they are suffering mutations in cancer-causing genes.

Another thing that happens with Werner's syndrome is that the tips of the chromosomes are worn off faster. In all people, the two tips of the chromosomes, known as telomeres, come together to form a loop that protects the DNA molecule. Each time the cell divides, the enzyme that does the copying can't quite turn the sharp bend in the loop, and a little bit of DNA is lost. Fortunately, the telomeres don't contain any genes, so all that is lost is a little "nonsense" DNA that doesn't mean anything. If the telomeres are seen as bumpers that eventually wear down to nothing, the problem comes when the telomeres can't get any shorter: the cell stops dividing. It doesn't actually die, it just stops growing.

The shrinking telomeres can be seen easily when human skin cells are cultivated in a lab. Cells taken from an infant have long telomeres. They go through about 50 divisions, which is roughly the equivalent of middle age, when the telomeres hit their critical length, and the culture stops growing. If the cells are taken from an older person with shorter telomeres, they will divide only a few times. It's as if the cells use their telomeres to know how old they are.

This process is the reason why young flesh is smooth and supple, while old skin is leathery and wrinkled. What makes skin supple is collagen, which is produced in large amounts by growing skin cells. When a person is young, damaged skin cells can divide to replace themselves. But with age, as the telomeres wear down to the nubs, the skin cells can no longer divide and actually begin to pump out an enzyme that destroys collagen. The ultraviolet rays in sunlight speed up the process by damaging skin cells and causing them to divide faster. Since each person has only 50 divisions to use during the course of a lifetime, the sooner they occur, the sooner the skin looks old. So be good to your telomeres, because they don't care about years, only about cell divisions.

In Werner's syndrome, telomere shortening is quickened. The lack of helicase somehow causes the tips of the chromosomes to be lost at an accelerated rate, just as the cells and organs are prematurely aging. The effect is even more dramatic in another extremely rare disease called Hutchinson-Gilford syndrome, which has only about 30 known sufferers in the world. In these cases, the skin wrinkles so quickly that children appear to be octogenarians, the heart begins to fail, and bones become fragile. Victims usually die of old age before they are 20.

If the telomeres really are ticking off the time we have left, then perhaps life could be extended by preventing the telomeres from shortening. One way to do this would be to activate an enzyme called telomerase that adds back DNA sequences to telomeres. This enzyme is active in eggs and sperm, which keeps those cells young and explains why older parents (with short telomeres) have children with youthful, long telomeres. Telomerase has been isolated and its gene has been cloned, so in principle it might be possible to extend the life span by activating the telomerase gene throughout the body. At least one biotechnology company already is pursuing this possibility, but there's a catch. Telomerase is produced not only in the germ cells of eggs and sperm, but also in cancer cells. The reason that cancer cells form tumors and metastasize is precisely because they don't know when to stop growing. So trying to prolong life by lengthening telomeres could actually have just the opposite result: the cancer cells could multiply, causing death.

Even if we were able to protect the telomeres and thereby prolong the life of cells in the body, there is one organ that wouldn't benefit: the brain. The wiring in the brain comes from neurons, and neurons don't lose their telomeres because they rarely divide. This does not protect neurons from damage, however, and they often are the first cells to fall to the aging process. In these cases, the body can remain perfectly healthy as the brain corrodes.

———

One morning, Martin was standing at the sink. Uncharacteristically, he hadn't showered or shaved. His clothes appeared wrinkled. Beatrice came into the kitchen to prepare his breakfast, as she had done every morning for 50 years. Martin looked at her and seemed confused. "Who are you?" he said. "What are you doing here?"

Martin was 74, and his condition began to deteriorate almost daily. His behavior became erratic, he often lost his balance and fell, and his blood-sugar level fluctuated wildly. Dangerously close to going into a diabetic coma, he was rushed to the hospital. When he was stabilized, he was sent to a nursing home, and Beatrice was told to stay away for three agonizing weeks while he became adjusted. "It was the hardest thing we ever did to put him in that nursing home," Beatrice said, her eyes filling with tears and her voice breaking. "I didn't want him to die away from his home."

A specialist examined Martin and called Beatrice from the home. "Physically, he is pretty good," he explained. "He's strong and in good health. I have to tell you, though, he is gone. You don't have a husband anymore. You have to know this so you can get on with your life. The mind is gone and the body will follow because the mind tells the body what to do. The mind tells you to eat, to go to the bathroom, how to walk. Everything is going to go."

The specialist was exactly right. At age 77, Martin was completely unable to function, and he died. The mourners at the funeral, held just a few blocks from the home where Martin and Beatrice had spent their married lives and raised two children, filled several rooms. His young grandchildren stood at the entrance to greet the arriving visitors.

THE AGING BRAIN

The story of Martin's quick decline will sound familiar. As some people age, their minds outlast their bodies and they remain mentally agile until the end. For many, however, advanced age seems to make them forgetful, distracted, and unable to focus. They can remember something that happened 50 years ago, but not 15 minutes ago. Memories become jumbled, thoughts confused. Words are lost and reading comprehension drops. This process was once thought to be a normal, inevitable part of aging and was called senility. Today, many of these people, including Martin, are diagnosed with Alzheimer's disease.

Four million Americans have Alzheimer's, the fourth leading cause of death in the United States. The disease was discovered in 1907 when Alois Alzheimer, a German physician, cut open the brains of people with dementia and found they were different from the brains of people who were still mentally healthy when they died. Just by looking at the outside of the brain, Alzheimer saw obvious differences. The sulci, the spaces between the folds of gray matter, were much larger in the troubled brains, indicating loss of brain tissue. This was especially true of the hippocampus, the repository of memories, and of the neocortex, the control center. The precise areas of the brain used for remembering and thinking were disappearing.

When Alois Alzheimer took out his microscope, he discovered the two hallmarks that remain the only sure diagnostic signs of the disease: plaques and tangles. The plaques are small circular areas, dense deposits of a protein called beta-amyloid, that resemble tarnished pennies lying in the sand. Surrounding these plaques are tangled webs of degenerating brain cells. In a healthy brain, the axons and dendrites of the neurons are elegantly extended to form the intricate wiring of the brain circuitry, like rigging on a sailing ship, but in Alzheimer's disease the lines are knotted, twisted, and snarled.

The beta-amyloid plaques are strikingly similar to the plaques seen in the brains of adults with Down's syndrome, a form of mental retardation caused by having three instead of the normal two copies of chromosome 21. This was an obvious clue that led scientists to investigate whether Alzheimer's might be caused by a gene on chromosome 21. Peter St. George-Hyslop and his colleagues at the University of Toronto focused on an unusual family that had a large number of people with Alzheimer's, and who were coming down with the disease relatively early, in their fifties. The researchers found a link to chromosome 21 and isolated a gene for amyloid precursor protein, which is chopped up to produce the beta-amyloid of the telltale brain plaques. The sufferers all shared a single genetic mutation that increased production of beta-amyloid and damaged the brain with lesions.

In this one family, at least, there appeared to be a simple cause of Alzheimer's. But when other sufferers were checked, the large majority of them didn't have a problem with the gene. In the end, only about 2 to 3 percent of familial Alzheimer's patients had the plaque-forming mutation on chromosome 21. Something else was eating up the other brains.

The gene hunt expanded to include other early-onset Alzheimer's families, including a genetically isolated group of Volga Germans who are valuable research subjects because they can trace their ancestry back to two villages in Russia. Soon two more suspect genes, called presenillin 1 and presenillin 2, were found on chromosomes 14 and 1. When the presenillin genes were put to work in skin cells, they caused the same overproduction of beta-amyloid plaque as in the earlier victims. This seemed to confirm that beta-amyloid is actually a cause of the disease and not a byproduct. But the numbers still didn't add up. There were too many cases of the disease in people who didn't have the mutations. All three genes accounted for less than 10 percent of Alzheimer's cases.

Then science was treated to one of those lucky coincidences that lead to breakthroughs. The two right people found themselves in the right place, and in the right frame of mind. The place was Duke University, where Margaret Pericak-Vance was a geneticist mapping genes for late-onset Alzheimer's, and Allen Roses was a biochemist studying the chemical composition of the senile plaques in Alzheimer's brains.

Pericak-Vance was looking at a type of Alzheimer's that is less obviously clustered in families, and at first she couldn't spot anything unusual in their DNA. But when she focused on pairs of siblings, she found a subtle link to chromosome 19. Most other geneticists didn't pay much attention, though, because she didn't show that everybody with that genetic code had the disease. Besides, there was nothing on chromosome 19 that looked even remotely related to the brain. In fact, the only thing in the neighborhood was apolipoprotein E, a fat-carrying blood protein that seemed unlikely to have anything to do with Alzheimer's disease.

In an amazing coincidence, Roses also stumbled across apolipoprotein E. He was looking for proteins that stick to the plaque-causing beta-amyloid, and one that kept coming up was apolipoprotein E. Roses didn't think much of it, since a lot of things stick to beta-amyloid, but when he saw Peri-cak-Vance's data a bell went off. They decided to work to-gether to see exactly what type of apolipoprotein E was pres-ent in Alzheimer's patients.

The gene has three different flavors or alleles, called E2, E3, and E4, which code for three slightly different forms of the protein called apoE2, apoE3, and apoE4. The researchers found that everybody—healthy subjects and people with Alzheimer's—could have any of the three types, but the peo-ple with Alzheimer's were more likely to have apoE4. Not only that, having two copies of apoE4 meant getting the dis-ease earlier. Having one copy of the gene, as do 27 percent of Americans, meant a three- to fivefold increase in the risk of getting Alzheimer's. Inheriting two copies, like 2 to 4 percent of the population, meant an eightfold increase in the chance of the disease and a 90 percent probability of Alzheimer's by age 90.

The mystifying thing about the finding is that it is so solid and yet it explains so little. There is no doubt this gene influences the onset of Alzheimer's; it's one of the best docu-mented results in all behavior genetics and has been repli-cated in more than 50 studies of people from around the world. Roughly 40 percent of Alzheimer's cases have been linked to this gene. Yet a link is not the same as a cause. Many people with the "bad" form of the gene never get Alzheimer's, and some people who get the disease don't have the gene. This also is true of other genes linked to behaviors; the evidence abounds but the explanations are elusive.

It's not even clear how a fat-carrying protein found mainly in the blood could lead to a brain disease. Roses's group has discovered that cells in the brain also produce apoE and that it binds to a protein found in the tangles surrounding the plaques. Current thinking is that apoE somehow facilitates the assembly of beta-amyloid into long, stringy molecules that pack densely and disturb the brain.

Someday genetic research may lead to new drugs to prevent and treat the disease. So far most of the drugs used to treat Alzheimer's attempt to restore neurotransmitters that are lost as a result of brain degeneration. They aren't very effective because they are trying to patch up the damage after it's done. It would be more effective to go after the source of the problem, the buildup of beta-amyloid into plaque. Until then, antiinflammatory drugs such as ibuprofin have some protective effect, probably by minimizing inflammatory responses to beta-amyloid in the brain. One thing that can be protected against is the risk of stroke, a major risk factor for Alzheimer's patients. Since the main risks for stroke are clogging of the arteries and high blood pressure, possible preventive measures include exercise, diet, and drugs for the heart and circulation.

An Age-Old Dilemma

Even if you escape dramatic genetic problems like Alzheimer's and Werner's syndrome, your cells are getting older, inevitably and irreversibly. We've long spent fortunes trying to stop the appearance of aging, to keep our hair from falling out or turning gray, and to keep our skin looking soft and smooth. Now we can't seem to resist trying to stop the aging process itself with dubious treatments, such as hormones like DHEA or melatonin. Although there haven't been enough clinical studies to show that any of the hormone treatments actually work, it's wise to remember that the original experiments with melatonin were on mice that were genetically incapable of making their own melatonin. When the experiments were repeated in normal mice, the only effect of the drug was to *shorten* life span by causing cancer. The problem with patching with hormones is that it's like trying to fix a leaky roof by drying out the rug. Because the roof hasn't stopped leaking, the rug is going to get soaked again.

We already have greatly increased our life span, not through changing our nature but by improving our nurture. Medical advances and improvements in living conditions are likely to continue to prolong life—or at least stave off death. We are creating a new set of problems, though, by changing the age balance of the population. For most of our history, the species has been young; the birth rate was high and death came early. Now we are slowing the birth rate and prolonging life, which means the population is getting older and more people need to be cared for by the young.

The other issue is whether we are adding years to our life rather than life to our years. Perhaps one purpose of a finite life span is to focus our attention. We know roughly how long we will live, and we plan our lives accordingly. There are bursts of energy to acquire knowledge during youth, then a rush to reproduce, followed by a collection of resources for retirement, and finally an age when we stop working. What if life lasted forever? Would there be an incentive to do anything productive? Maybe, but nothing gets people moving like a deadline.

THE SOUL GENE

Not everything about getting old is depressing, however, and there is some proof to the saying, "You're not getting older, you're getting better." While underlying temperament is not likely to go away as you age, character continues to change and often for the better.

Temperament is the most fixed part of personality. For example, when Robert Cloninger looked at 1,019 people from age 18 to 99, he found that harm avoidance, a measure of anxiety and worry, was not at all related to age. People didn't seem any more or less anxiety-prone whether they were older or younger. This trait, which colors the whole view of life and is in part mediated by serotonin genes, appears to be stable during an entire lifetime. Likewise, reward dependence, which measures the need for approval, is constant across age groups. Novelty seeking does decline with age, which is why teenagers pay extra for car insurance, but young thrill seekers grow up into adults who will always be relatively more interested in new sensations. The outlet might change from driving fast to becoming a police officer, but the desire for stimulation doesn't disappear.

The best hope to modify temperament is up to age 30 or so. In some of the most careful and extensive studies of this type, Paul Costa and Robert McCrae at the National Institute on Aging examined personality in four different cultures. In all four cases, they found several changes between ages 18 and 30, including a decline in neuroticism and increases in agreeableness and conscientiousness. During this "settling in" period people become less distressed, more social, and better organized; in other words, more mature. But after age 30 there is remarkably little change in basic, genetic personality traits. For example, the best way to predict a person's happiness at age 80 is not to ask about health, wealth, or loved ones; it is simply to look at the happiness level 50 years before. Says McCrae, "If you're cheerful, well adjusted, and methodical at age 30 you will probably be cheerful, well adjusted and methodical at age 80. Of course, if you're gloomy, anxious, and sloppy at age 30, the prospect isn't so good."

Even though underlying temperament is not likely to change, people do learn from experience. Character, the acquired part of personality, can improve even at an advanced age. Cloninger found that as people move from young adulthood to the last decades of life, they become more willing to help others and to improve themselves. His scales for helpfulness, compassion, and conscience all show consistent gains throughout middle age into older age. People become more willing to help others, and less self-centered and egotistical. They also become more forgiving and less likely to seek revenge. This is especially useful in relationships, as people become more open to overcoming difficulties and less likely to fight with their partners. People become more honest, sincere, and willing to treat others fairly, rather than selfishly pursuing their own goals. It's not that they lose ambition, but they are more scrupulous in how they achieve it.

People also can learn to deliberately change behavior. What starts as self-discipline or will power, such as an angry person who practices counting to ten, can become a new ingrained behavior. This is the way an alcoholic keeps out of bars and a smoker gives up cigarettes for good. Each time we exercise will power, we rewire the brain to overcome inborn temperaments. Many bad habits result from inherited temperament, such as anger, depression, impulsiveness, and addiction. Choosing good habits takes hard work, but the sharp edge of temptation can be dulled with practice. The longer we practice good behavior, the easier it becomes, until it becomes habit.

Cloninger found one other thing that seems to increase with aging: spirituality. Not only that, but spiritual people are relatively more likely to express warmth, altruism, positive emotions, and openness to feelings. They are relatively more intimate and friendly, they are generous, and concerned for the welfare of others. They also tend to be more optimistic about life and more likely to experience positive emotions such as love and happiness. Above all, spiritual people are open to their own inner feelings and emotions. They experience happiness and joy—as well as pain and suffering—with heightened intensity.

The essence of spirituality, which can include a belief in God or higher power or a divine order to the universe, is looking inward, searching for meaning and purpose, and seeking to understand what truly matters. People turn away from materialism in search of inner peace, through identification with God or with the cosmos. Is spirituality simply an adaptive response, a self-deception to deal with old age, infirmity, and death? Or is it wisdom, a gradual realization of the real truth about the universe? A scientist might wonder whether spirituality wasn't written into our genetic code. Perhaps as the body begins to expire, the brain wires a new set of neuronal connections in the cerebral cortex that allows us to accept the end with grace, dignity, and even hope. On the other hand, this may be a lot of mumbo jumbo, a far too clinical explanation for what we know as the soul.

ENGINEERING TEMPERAMENT

Cloning and the Future Politics
of Personality

Andrew thought a lot about his father, if that's what you could call him. Andrew's father was a scientist who had left a respectable career at a quiet university to join a risky bio-tech venture in the year 2002. The risk paid off, big time, and Andrew's father became a multimillionaire. More importantly, he was able to work on the project of his dreams: cloning. The startup firm, Mirror Image, Inc., worked on the genetic cloning of animals, including humans. Wall Street loved the idea, and the stock's initial public offering set unheard-of records. Investors bet heavily on the company because the technology was relatively simple and the promise enormous. The only possible roadblock was Washington, and there was the risk that the government would try to ban cloning research. There had been congressional hearings and public handwringing after the cloning of the first sheep, named Dolly, but the concern had eased. Scientists had showed that animal cloning could help feed the world, speed medical advances, and provide other benefits. Cloning had become less scary and more acceptable.

Human cloning was a little more complicated, but only in terms of ethics and politics. The science was the same. The ethical questions were still being debated, of course, just like people still debated how many angels could fit on the head of a pin. The more practical minded people were simply going ahead and doing it. Mirror Image pursued a two-track policy. Publicly, company officials said little, to avoid scrutiny. They weren't doing anything illegal, in fact company policy was to follow the highest legal standards to avoid being tripped up by a technicality. But behind the scenes, they were funding research at powerful think tanks and courting policymakers. Changes in the campaign finance system back in the 1990s had not worked exactly as the reformers had planned, and Mirror Image and similar companies were able to quietly influence appointments to the congressional committees overseeing their industry.

The company was doing the science openly, on animals that is. Secretly, the scientists worked on human experiments, slightly altering the protocol that had been used with sheep to make it suitable, tweaking the cell growth media, testing various methods of preparing nuclei, etc. Not actually cloning any humans, yet, but doing all the tedious but necessary preliminary experiments. In case of an emergency, most likely of the political kind, the company had built an exact replica of its U.S. lab in a secret location overseas.

Andrew's father wasn't much interested in the politics or the ethics. He was an intensely focused man interested in technique and results. He wasn't naive, however, and he knew that he had to be careful. He kept a detailed record of his experiments, his lab was meticulously run, and his public behavior was exemplary. Every step was planned and projected, and recorded on videotape, when he inserted a swab into his mouth, scraped a few cells from the inside of his cheek, and dabbed them onto a petri plate. He began the cloning process.

When his cells had progressed to the ideal stage, they were treated with a chemical that arrested growth, putting them in a temporary state of suspended animation. Then the nucleus from one cell was removed and inserted into a human egg, donated by a lab assistant, that was in the same suspended state. The DNA from the egg was replaced by the DNA from the cheek scraping, but the egg had no way of knowing this. As far as the egg was concerned it had just been fertilized in the normal way, and so it began to divide and to copy the scientist's DNA. Soon the egg was reimplanted into the womb of the lab assistant, and nine months later, a bright and healthy baby boy was delivered into the world. His name was Andrew. His DNA was identical to that of the scientist. Not 50 percent the same, as would be the case for a normal child, but base-for-base the same, as if Andrew and his father were identical twins.

The birth announcement was greeted with shock and anger. While everyone knew that it was possible to clone a human, technically at least, no one expected that it actually would be done. Anticipating the reaction, Mirror Image had prepared a strong defensive strategy, and it paid off. After the initial outrage, with magazine covers comparing Andrew's father to a Nazi scientist, the coverage began to switch focus to the baby. The baby was innocent of any wrongdoing and was undeniably cute. He cooed and gurgled at the TV cameras and seemed to love the attention. Network producers fought for the rights to chronicle his every move, and Andrew became a kind of national pet. "Andrew Eats Solid Food," trumpeted the *New York Times*. A tabloid reported, "Nothing Sheepish About Our Andy!"

The coverage began to fade after a few years, mostly because Andrew did nothing extraordinary. There were only so many stories that could be done about a normal, healthy toddler. By the time he was five, his first day at school was mentioned only in passing by the major news media. He was just another kindergartner. Andrew was raised in the modest suburban home of the lab assistant, who couldn't have loved him more. She stopped working while he was small to be with him, and Mirror Image provided more than enough money to support the little family of two. Andrew's father saw him only when the boy came to the lab for tests. The father had moved on to other experiments, and like the rest of the nation, he didn't much care about Andrew after the novelty had worn off.

The only person who remained obsessed with Andrew's story was Andrew. He had indexed all the newspaper and magazine stories about his birth, as well as the scientific abstracts. Interestingly, when his father was younger he had made a similar scrapbook, collecting thousands of articles about DNA and cloning, even though in those days it was just cloning of carrots and frogs. Andrew built his massive scrapbook on a computer with all the video and digital images of his life. He used his "Clone at Home" software, a hugely popular program, to replay his creation and development. The software allowed a person to plug in different variables, genetic or environmental, to his or her life story. Then the computer predicted how the life would have developed, and a video image showed how the person would age from birth to 80 years old. Andrew's favorite game was to delete his father's DNA and replace it with the DNA of his "mother," the lab assistant. When Andrew played out the program with her genes, he preferred the virtual results to the real thing.

It was too late to delete his father's DNA, however. All of Andrew's potential, all that he ever would be, was contained in that one swab dragged across the inside of a cheek. Andrew stared at himself in the mirror and tried to see himself clearly. Who was he? He knew he was a separate person, but it was all so confusing. When he compared himself with old snapshots taken when his father was a boy, he could hardly tell the difference, for they were as alike as identical twins. And yet he didn't *feel* like his father, or at least he didn't think so. The resemblance was so eerie that he deliberately wore his hair differently from his father and experimented with pierced ears, which didn't suit him. In his face, Andrew also noticed traces of his grandfather and grandmother. Andrew did some research on the family of his father and fell in love with his great-great-grandmother, a pioneering bioethicist who had warned, unsuccessfully as it turned out, about the dangers of genetic engineering. Andrew idolized her and saw her as his true forbearer, not the man who scraped a cheek. Andrew came to see that man as merely one disposable receptacle for the long march of his genes, with about as much personality as a beaker.

Andrew could not deny, however, that there were elements of his own personality that came from his father. They shared a sense of being proper and neat; it infuriated them both that the lab assistant allowed her car to fill up with old newspapers, gum wrappers, and empty Coke cans that rolled back and forth under the seat. They both had short fuses. Andrew's father tended to blow up when things went wrong;

he once threw a batch of improperly prepared gel at a cowering postdoc. Andrew also had his rages, but he controlled his anger with exercise; when he felt the pressure building, he'd go to the gym or take a run. Andrew accepted those parts of his personality as givens, unchangeable but controllable. What he did with his life was another thing; that was his choice alone. He wasn't going to follow in his father's footsteps. In fact, he vowed to use that same focused determination and intellect that made his father such a good scientist to stop the people who believed that cloning humans was no different than mass-producing sheep.

Is Andrew's story farfetched? A little, but only to make a point. Both the knowledge of human genetics, and the ability to manipulate and exploit it, are expanding at an exponential pace. By the first decade of the twenty-first century we will know the entire sequence of the human genome, every one of the more than three billion nucleotides that make up the 100,000 genes that constitute our genetic patrimony. Deciphering the meaning and function of those genes will be slower, but success is inevitable. Already, an entire new field has emerged called "functional genomics" dedicated to figuring out what genes do. At the same time, new technologies are emerging to exploit this information with drugs and by manipulating the genetic information itself. So far the manipulations are being performed only on animals, Dolly the sheep being the first well-known example, but humans are just a few steps away.

There is no turning back. Supporters of the international Human Genome Project argue convincingly that mapping the entire genome will help produce new drugs, reduce birth defects, and allow us to live longer and healthier lives. Even if the project were officially stopped, much of the most advanced, demanding research has been taken over by biotechnology start-up companies, and by the giant pharmaceutical firms that usually end up owning them. Lives are at stake. Money is at stake. That's a powerful combination anywhere, and in America it's invincible.

The focus of attention so far has been on discoveries of genes for cancer and other physical illnesses. What often goes unsaid is that the genes being discovered also include ones that define behavior. Virtually every aspect of how we act and feel that has been studied in twins shows genetic influence, and many of the individual genes have been isolated. The fact that so many behavior genes are being found is not surprising because the brain is so complex that much of our genetic information is devoted to building, developing, and maintaining it.

The combination of these two forces—the stampede to map the genome plus the decisive role of genes in behavior—means that, whether anyone thinks it's a good idea or not, we soon will have the ability to change and manipulate human behavior through genetics.

As scary as that might sound, the genetic manipulation of behavior is not new. Humans have been selectively breeding behavior in animals, such as dogs and farm animals, since before history. And whether we recognize it or not, we already are products of genetic engineering. As far back as we can see, humans have been selective about their mates. Beauty, power, and prestige have always been desirable. Clans or families sought to marry their offspring with people of high rank and stature. People from particular religions or races sought their own kind. The highest social classes, the blue bloods, sought to preserve the purity of their lines.

The difference is that in the near future science will give us the power to do it more quickly, more accurately, and more decisively. We will select mates not because of some superficial trait like a prominent family; we'll be able to read their DNA as easily as an X-ray. Nor will we settle for the randomness of sexual blending of sperm and egg, with all its billions of possible combinations, when we can build the desired combination to the letter.

It's too late to wonder whether we are going to genetically tinker with human behavior. We need to decide very quickly how we are going to do it. How will we distinguish "good" genes from "bad?" What traits will be valued and what will be discarded? Who gets to choose?

DNA on a Chip

The first stage of the process will be using genes for diagnosis, rather than for manipulation. Currently, most mental health experts analyze personality using very unsophisticated tools: their eyes and ears. They listen to patients, compare their symptoms with those of recognized syndromes, and make a "diagnosis." Only it isn't a diagnosis at all because it doesn't say anything about the underlying mechanism or origins of the problem. It's just a description: a matching of the patient's complaints to complaints that other people have described.

In the future, a person who complains of depression or anxiety could have a DNA test to check the serotonin genes.

People with compulsive behavior such as gambling, drinking, drugs, or promiscuous sex, would be checked for dopamine genes. Eating disorder or obesity? Look at the genes for leptin, the leptin receptor, and its targets. A new technology called DNA chips, already under development by a biotech firm called Afymatrix, will make an entire DNA blueprint as easy to read as a supermarket bar code.

Doctors won't be the only ones to read this information. Insurance companies, which profit by charging based on risk factors such as smoking, would be very interested in genetic predispositions toward addiction or mental disorders. The military, which today rejects people who took medication as teenagers to control attention deficit disorder, might want to know about genes for rebellious temperament. Employers might be interested in genes for loyalty. Religious orders would be wise to discourage high novelty seekers, while the maker of sports cars would want to target them with ads. Dating services would have revealing new ways to match people. Imagine how excited certain school administrators would be to track students who are bright, troubled, or aggressive.

We will have all new ways to understand people—and to label them, discriminate against them, or help them. The technology is coming; how to use it is up to us.

DESIGNER DRUGS

The second phase will be designer drugs tailored to individual genetic needs. Today's pharmacology is so imprecise that doctors diagnose by prescribing drugs; if Prozac works, the patient must have a serotonin problem. Why not start backwards and build a drug that addresses the problem? Individual DNA sequences, combined with existing computer modeling techniques, would allow drugs to be as precise as a key in a lock.

The customized brain drugs will certainly be an improvement over today's sledgehammer method of pounding the body with chemical combinations until something works. They also will create new dangers, however. For one, we know nothing about the consequences of changing brain chemistry. The brain is so sensitive that it might respond to our manipulations in unforeseen ways. Already people are taking drugs such as Prozac for years or decades, even though the initial safety tests lasted only months. What if the brain's long-term response to the drug is to change the wiring pattern in a way that true joy, exuberance, or ecstasy were impossible? Also, imagine the possibilities of selling illegal pleasure drugs, custom-made to individual brains.

Another danger is that we will medicalize normal human behavior and variations. Some people argue that this already is happening, that Prozac is being prescribed for normal mood swings, and that Ritalin is being used to "treat" boyhood. If drugs designed to fight life-threatening obesity are used casually to drop a few pounds before prom night, why not use a behavior drug to improve your own particular mood? Why should you ever feel blue if you don't have to? Why should teachers have to tolerate disruptive behavior when they could spray kids with a calming aerosol that actually would help them learn? Maybe crime could be reduced by smoothing out the rough edges of problem youths. We wouldn't need assertiveness training if girls could take pills to cure shyness. What about that pesky gay gene? Spray it away with new "Straight-in-a-Day!"

GENE THERAPY

But why bother with pills or sprays? Why treat the symptoms if you can fix the cause? If all these mental "problems" are caused by genes, then why not change the genes themselves? This is the ultimate stage of the coming revolution: gene therapy, the deliberate manipulation of genetic makeup to repair a defective or mutant gene, to replace it with a "better" gene, or even to introduce a completely new gene.

One way of doing this would be to alter cells directly in the brain. New genes already can be delivered into cells by packaging them inside viruses, which are designed by nature to enter cells. Already there are hundreds of clinical trials underway using gene therapy to treat simple disorders caused by a single gene, such as cystic fibrosis. Much of the difficulty at this point has been getting the new genes into the cells, so scientists are looking for new genetic delivery vehicles. Inder Verma at the Salk Institute in La Jolla, California, is experimenting with the notoriously effective virus, HIV, which causes AIDS. He has managed to disarm the AIDS part of the virus and has enabled it to penetrate many types of cells. This feared virus could become the ultimate "suitcase" for carrying genes.

It's easy to imagine how gene therapy could be used on severe brain diseases such as Parkinson's and Alzheimer's. Since Parkinson's is caused by a loss of the cells that produce dopamine, it might be possible to insert a gene that results in dopamine production. In the case of Alzheimer's, in which neurons are lost because of the formation of plaques and tangles, it might be possible to induce new neuron growth by expressing the gene for nerve growth hormone, a sort of growth hormone for brain cells. The next likely target for brain gene therapy would be the severe psychiatric disorders such as schizophrenia, although everything would depend on determining the genetic blueprint of the disorders.

If gene therapy for schizophrenia worked, it would be logical to consider gene therapy for less severe conditions, such as depression or mental fatigue. If sufferers had the "wrong" form of the serotonin transporter gene, it theoretically could be swapped out for the more upbeat version. This would be an exaggerated treatment, though, since the same effect could be obtained by taking Prozac. There also could be undesirable consequences, such as the loss of sex drive. More importantly, gene therapy would be permanent. Unlike taking a drug, there would be no way to undo the treatment. A bigger concern is that genes for personality are so complicated that it's impossible to know what might happen if one were altered. The fabric woven by DNA in combination with the environment is so intricate that unravelling a single thread could destroy the whole tapestry.

Other problems with this kind of gene therapy on brain cells would arise because it usually would be used to clean up damage rather than prevent it, and it's good for only one generation. The best way to fix a gene before it is passed on to the next generation is to go into the germ cells in the sperm and eggs. One approach to this "germline" therapy has already been developed, at least in mice. New DNA is introduced into cells that are grown in a culture, then mixed together with natural cells from an early embryo. Some of the resulting babies will have the engineered genes in their germ cells and can pass them on to the next generation.

This all sounds straightforward enough, but the human genome is so large that when a new bit of DNA is introduced into a human cell, it only rarely figures out where it's supposed to go. Nobel Prize winner Paul Berg is experimenting with how to guarantee delivery and make single base changes in the mammalian genome. Berg learned from watching yeast that a certain break in a DNA molecule makes it much easier to recombine with engineered DNA. He has managed to break mouse DNA in a way that encourages it to incorporate any change that he desires, thus clearing the way for narrowly targeted genetic engineering.

Berg, however, scoffs at the idea of genetically engineered humans, not because it won't become possible but because he doubts people will want to. His own hands-on experience with human engineering occurred when he was approached by a stranger carrying a little black bag. The man said he was collecting sperm from Nobel Prize winners and wanted a sample from Berg, which then would be sold. This is an obvious kind of genetic engineering, and there's no reason it can't work, except for the fact that people simply don't want to do it. Berg sent the man packing, telling him with a laugh that the scheme would flop. "I know a lot of Nobel laureates," he told the sperm banker. "And I know their children."

THE UNCERTAINTY PRINCIPLE

Even if all the formidable technical hurdles to genetic engineering were cleared, there would remain a deep, fundamental difficulty because of the very nature of genetic influences on behavior—the uncertainty principle.

As shown time and again in this book, genes may predispose toward a particular type of temperament or behavior, but they are never sure predictors. A person may have genes that predispose him to addictive behaviors like alcoholism, but he could just as well become an avid bug collector who discovers an important new species. Another person might have the genetic makeup typical of a mass murderer; but he could turn out to be the next great professional linebacker.

The problem with germline therapy is that it targets genetic "problems" before they have manifested themselves. If left alone, traits that could be regarded as problems might never develop, or they could turn out to be assets. If it's true that there is a fine line between genius and madness, or between creativity and depression, then we should be careful what we engineer for because we just might get it. A DNA map is not the same as a road map, which shows exactly how to get from one place to the next. A DNA map offers possibilities and predictions but not certainty.

The uncertainty principle also includes unintended consequences. Even if we know exactly what a gene is for, say baldness, that doesn't mean that neutralizing the gene will cure baldness. Even it did cure baldness, the gene might also be involved in vision or smell or some other function that we could never have imagined. Consider, for example, what happened when scientists used gene surgery to improve the muscle mass in mice, research that could be of great benefit to people suffering from muscle disorders.

A gene called myostatin controls muscle growth. When the body has the normal amount of muscle, the myostatin prevents the muscle cells from growing any further. Scientists easily knocked out the gene and bred a strain of mice without myostatin, who therefore had no block on muscle production. By six weeks, the mice had developed unusually big shoulders and hips. When they were full grown, their skeletal muscles were two to three times larger than normal mice.

Other types of muscle, such as in the heart and intestines, were not changed. The scientists speculated that this type of gene therapy could be used to make leaner, meatier farm animals. Some day it could also be used on humans to treat the loss of muscle that occurs in muscular dystrophy and in some types of cancer and in AIDS.

But there was another, totally unexpected result: the muscular mice were wimps, what the scientists politely referred to as "gentler." The big mice were less eager to attack other mice and less likely to react when poked or prodded by human handlers. When Alexandra McPherron, one of the scientists at Johns Hopkins University who developed the new strain, presented the mice to the world at a news conference, the mice started fighting. McPherron commented matter-of-factly, "The normal one there is beating up the big muscular one."

No one expected a personality change. The only effect on the mice was supposed to be in muscle mass, not in the brain. Everybody thought they knew exactly how myostatin works on muscle growth, and in fact they did. The mice did get bigger, but the gene must have some other still unknown role. Or perhaps taking out that gene changed the expression of another gene, or a hormone, or something else in the body that altered personality. There could be 10,000 steps between muscle mass and a passive personality, but when 1 step is removed the entire developmental process shifts.

KNOW THYSELF

Despite the risks of trying to predict what genes will do, there are ways to benefit from understanding your own genetic makeup. For example, if your family has a history of heart disease or breast cancer, it would be wise to have regular checkups and lead a healthy lifestyle. If obesity is a family problem, it would make sense to develop good eating habits early in life. If your father and brothers are alcoholics, you might be better off not drinking. Those are all commonplace examples of preventive measures based solely on genetic information. They reflect the popular wisdom that physical health is in part a product of genes. The only difference with genes for personality and behavior is that the effects are more subtle and still less understood.

When you think about the genes for your own personality, you are asking, "Who am I?" The simple answer is, "You are who your brain thinks you are." And who your brain thinks you are is the result of an intricate, one-of-a-kind interaction of genes and life experiences.

In the very first days of life, perhaps from the moment of conception, a baby is learning about itself and its world. How does a baby come to know its own body? How does it know where its toes are, and how many it has? How does it distinguish between a mother's caress and a slap on the face? All the touch-sensitive cells in the body make connections through the central nervous system, up the spinal cord to the thalamus and the brain's somatosensory cortex. There the brain cells are arranged in the form of a homunculus—a "tiny man." This little man is created in our image, sort of. He's upside down, so that neurons that respond to stimulation of the face are at the bottom, and those tied to the toes are on top. And he's distorted: the regions representing the face, hands, and genitals—the most touch-sensitive regions—are disproportionately large.

If you could open up the brain and probe it with electrodes, you would find that stimulating the cells at the top of the homunculus would "feel" like someone massaging your toes, or that touching the bottom of the strip would make you "feel" something on the tip of your tongue. This is because physical feelings are nothing more than the firing of specific brain cells.

We know three important facts about the homunculus. First, its ability to develop is genetically hardwired. Imagine the confusion if it weren't: if you were born with a "little dog" up there instead of a little man you'd be barking for your dinner. Second, the proper formation of the homunculus is dependent upon experience; although the genetic instructions are necessary, they are not sufficient. For example, if a rat is shaved of all but one whisker on its snout, and therefore can't receive the expected sensations from the other whiskers, the rat cortex develops a huge bunch of cells to receive information from the one remaining whisker. The areas tied to the missing whiskers disappear from lack of use. Third, the somatosensory cortex can change throughout adult life. For example, practicing a particular violin passage over and over leads to the expansion of the neurons that receive the signal from the relevant fingers. By practicing the violin passage, you can change how the brain is shaped and how it works. Although the DNA had a general blueprint for the homunculus at conception, it is not averse to changing to meet a later need.

To understand how personality is shaped, imagine that the brain not only has a physical homunculus that registers touch but a temperamental one that registers emotions. Each person could have an individual map of what makes him or her feel good or bad, anxious or relaxed, excited or complacent. We don't yet know much about where this hypothetical emotional homunculus is located, but it's probably more spread out than the somatosensory cortex. Nor do we understand how it varies in different people, but differences in both genetic makeup and experience are probably involved. Perhaps in one person the "rage" area is exaggerated and the "altruism" area is tiny (avoid this person). Perhaps in another person it is the excitement area that is large and the fear center that is small (don't get in her car). The empathy area could be large, which means this person feels good by helping others (marry him).

Under our theory, the temperamental homunculus develops from a combination of hardwired genetic instructions and experience. Most people are born with the capacity for the full range of human emotions: fear, anger, and sadness; joy, love, and lust; surprise, disgust, and shame. But those emotions can only develop in response to experience. Without tapping those feeling centers at the right time, they cannot develop. Just like in the rats with one whisker, the unused parts of the brain could disappear.

This is why attention to childhood development is so important. Critics of behavior genetics say it limits people by defining their lives in terms of inherited traits. That would be a mistake, just as it is a mistake to judge people by the color of their skin or place they were born. But the real tragedy is that so much genetic potential is never developed. How much greatness there is in the world that is lost because of inattention and a lack of love. How many beautiful, perfect little seeds are brought to life every day without the opportunity to grow to their strong and healthy potential. In our own country, we have eliminated many of the barriers to physical growth, such as disease, but the breakdown of the family makes it that much harder to help people reach their emotional potential.

On the other hand, the story of Nicholas Scoppetta, who rose out of a broken and troubled home to become a top attorney and New York's commissioner of child welfare, shows that people are remarkably resilient. Just as a perfect upbringing is no guarantee of success, a "bad" childhood is not a predictor of failure. In fact, maybe some people only develop all their natural instincts when challenged by a difficult environment. Perhaps adversity is what hones our genes to razor sharpness. In any case, it's clear that environment is not the only cause of personality problems. Even if we somehow managed to clean up all the slums and gave every child a hot breakfast, we would not have a perfect population. We still would fight, cheat, steal, kill, sleep late, and sleep around, drink too much, be petty, cranky, and obnoxious.

As humans, we are a diverse mix of personalities, a jumble of traits so complex that we barely come to know ourselves in the course of a lifetime, let alone understand what makes other people tick. The exciting thing about the new science is that it is giving us another tool to understand ourselves. Not too long ago, people "went crazy." Today they have chemical imbalances that can be fixed easily. Suicidal depression was often untreatable. Today many people owe their lives to a simple chemical that helps their brains. The logical extension of these breakthroughs in brain chemistry is understanding the genetic roots of personality. Genes aren't scary; they are a fact of life. Gene therapy isn't scary, either, although there are risks. There are similar risks in other scientific advances, however, such as the possibility that our reliance on life-saving antibiotics will produce new strains of superbacteria. Would anybody argue that we should never have developed antibiotics? The greatest risk has always been ignorance.

So what about free will? It's alive and well, and probably genetic. Free will means taking control of your life. This is only possible when you understand who you are. As humans, we are born with instincts to survive, to love, to reproduce. As individuals we are born unique, each of us a variation on the human theme. Genes play an essential role in the overall theme and the individual variations; genes make us human and they make us unique. People cannot be mass produced, even if we tried. Every individual has too many choices and too many possibilities to ever predict the future. You are born with a pen and paper in hand, but you have to write your own story.

Andrew refused a college scholarship from Mirror Image and worked at a restaurant to put himself through school. He studied philosophy and ethics, and wrote his Ph.D. dissertation on the work of his great-great-grandmother. There wasn't much money in philosophy, but he landed a job at a small think tank in Washington. The director remembered the story of Andrew's birth, and thought it might make a good hook to pull in donations for the center. He set Andrew working on the ethical issues of genetics, a perfect fit for the think tank and for Andrew. One of the major backers of the center, however, was not amused. Mirror Image had endowed the center before Andrew's birth, and it's mission was to provide ethical protection for gene work, not question it. Andrew was soon out of a job.

There was a flurry of publicity over the firing and a

series of retrospective stories about Andrew's life. The media, of course, found it fascinating that Andrew would rebel against the very technology that created him. But people with a slightly longer view realized there was nothing unusual about a boy rebelling against his father. Andrew's father had the unique experience of being rebelled against by himself. In interviews he joked about having a split personality and said he was "of two minds" about Andrew's progress. Not everybody got the joke.

Andrew knew from his own experience that the attention wouldn't last, so he had to act fast. He set up his own foundation, attracted a few major backers, and began a campaign that would lead to the first effective legislation controlling the gene industry. He became a familiar figure on the talk shows and an articulate defender of the uniqueness of the human spirit and the power of free will. He kept alive the age old questions that had been pushed aside in the excitement over cloning: Who are we? Where do we come from? What should we do with our lives?

Andrew married a woman he met in college, and they had two beautiful children. Andrew and his wife were thoroughly modern and dutiful parents, reading to their babies and playing classical music in the nursery. Before the children were conceived, Andrew and his wife had gone to their obstetrician/geneticist for routine counseling. The doctor had offered to run their DNA through the scanner and discuss any modifications they might want to consider. Andrew, of course, was opposed to genetic engineering, but they were a little concerned because Andrew's wife had a brother who was a paranoid schizophrenic, a condition with strong genetic roots. The doctor assured them that routine screening would weed out such obvious flaws in the DNA. Severe mental conditions, obesity, extreme hyperactivity, socially limiting shyness (SLS), and criminal aggression all had been virtually eliminated through routine genetic screening before conception. People no longer aborted babies with obvious problems; such babies were no longer conceived.

Andrew and his wife decided to play it safe and have the screening. Whatever their own ethical views, it wasn't fair to the child to ignore the latest medical technology. The world was a harsh enough place that it wasn't right to doom a child to a lifetime of suffering when it could be prevented. All their friends did much more refined screening, looking for football players, concert pianists, mathematicians, or chefs. Just like certain baby names would come into style periodically, so did personality traits. The hot trait these days was insouciance, which was patented by a leading perfume maker. The issue wasn't whether to check the unborn DNA, it was what to check it for. Because of his own experience, Andrew had very strong feelings about this. On the DNA screening test, they checked "yes" next to the boxes for severe retardation and Extreme Personality Profile (EPP), which in most cases led to institutionalization. But they checked "no" in the sections for musical ability and poetry.

Andrew and his wife knew more about their children than any generation of parents before them. For the first time in history, they had both nature and nurture covered. They had read everything about childhood development, and they also had the raw genetic code. Andrew understood better than anyone what this information meant: very little. The wondrous truth was that he would have no more say in the future direction of his children than he would in the flight of a bird. All he could do was open the door of the cage and say, "Fly!"

NOTES

INTRODUCTION: EMOTIONAL INSTINCT

1. The story of the Jim twins and Amy and Beth was taken from Lawrence Wright, "Double Mystery," *The New Yorker,* August 7, 1995, 44–62.

CHAPTER ONE: THRILLS

1. For the sake of simplicity and uniformity, we use the term "novelty seeking" throughout this chapter, even though several of the experiments that are described actually used Zuckerman's Sensation-Seeking Scale or other measures such as the NEO five-factor inventory. Empirical studies have demonstrated a strong correlation between these different scales.

2. The strength or effect size of the association between a gene and a trait is frequently described using standard deviation units, which are

a standardized measure based on the population variability of the trait. The standard deviation is mathematically described as the square root of the average squared deviation from the mean. For personality traits, which follow a normal or bell shaped distribution, the standard deviation corresponds to that value of the trait such that 68 percent of all cases fall in the interval between one standard deviation above the mean and one standard deviation below the mean. Thus if a particular gene variant has an effect size of one standard deviation unit, individuals with one form of the gene will score on average 16 percent higher for the trait than individuals with the other form of the gene. An association of 0.5 standard deviations units indicates an 8 percent difference, and so on.

CHAPTER TWO: WORRY

1. The population frequency of the long version of the serotonin transporter gene is 57 percent, and the frequency of the short version is 43 percent. Therefore the expected frequency for long/long homozygotes is $0.57 \times 0.57 = 32$ percent, the expected frequency for long/short heterozygotes is $2 \times 0.57 \times 0.43 = 49$ percent, and the expected frequency for short/short homozygotes is $0.43 \times 0.43 = 18$ percent. In agreement with expectation, the observed gene frequencies for 1,010 chromosomes were 32 percent long/long, 49 percent long/short, and 19 percent short/short. The experimental data indicates that the short version of the gene is dominant to the long version both for expression and personality effects. Thus 32 percent of the population has higher serotonin transporter gene expression, and the remaining 49 percent + 19 percent = 68 percent has lower serotonin transporter gene expression.

CHAPTER SIX: THINKING

1. The story of Nicholas Scoppetta is taken from Dale Russakoff, "The Protector," *The New Yorker,* April 21, 1997, 58–71.

2. The Tonegawa experiments focused on the NMDA receptor, a form of the glutamate receptor that also recognizes N-methyl-D-aspartate.

SOURCES AND FURTHER READING

INTRODUCTION: EMOTIONAL INSTINCT

Bishop, Jerry E., and Michael Waldholz. *Genome.* New York: Touchstone, 1990.

Bouchard, T. J., Jr., D. T. Lykken, M. McGue, N. L. Segal, and A. Tellegen. "Sources of Human Psychological Differences: The Minnesota Study of Twins Reared Apart." *Science* 250, 223–8, 1990.

Bouchard, Thomas J., Jr. "Genes, Environment, and Personality." *Science* 264, 1700–1, 1994.

Buss, D. M. "Evolutionary Personality Psychology." *Annual Review of Psychology* 42, 459–91, 1991.

Carlson, Neil R. *Foundations of Physiological Psychology.* 3rd ed. Boston: Allyn and Bacon, 1995.

Cloninger, C. R. "A Systematic Method for Clinical Description and Classification of Personality Variants." *Archives of General Psychiatry* 44, 573–88, 1987.

———. "A Unified Biosocial Theory of Personality and Its Role in the Development of Anxiety States." *Psychiatric Development* 3, 167–226, 1986.

———. "Temperament and Personality." *Current Opinion in Neurobiology,* 1994.

Cloninger, C. R., D. M. Svrakic, and T. R. Przybeck. "A Psychobiological Model of Temperament and Character." *Archives of General Psychiatry* 50, 975–90, 1993.

Cloninger, C. Robert, Rolf Adolfsson, and Nenad M. Svrakic. "Mapping Genes for Human Personality." *Nature Genetics* 12, 3–4, 1996.

Cloninger, C. Robert, Thomas R. Przybeck, Dragan M. Svrakic, and Richard D. Wetzel. *The Temperament and Character Inventory (TCI): A Guide to Its Development and Use.* St. Louis, MO: Center for Psychobiology of Personality, 1994.

Costa, Paul T., Jr., and Robert R. McCrae. *Revised NEO Personality Inventory (NEO-PIR) and NEO Five-Factor Inventory (NEO-FFI): Professional Manual.* Odessa, FL: Psychological Assessment Resources, Inc., 1992.

Eysenck, H. J. *The Biological Basis of Personality.* Springfield, IL: Charles C. Thomas, 1967.

Gallagher, Winifred. *I.D.: How Heredity and Experience Make You Who You Are.* New York: Random House, 1996.

Goldberg, L. R. "An Alternative Description of Personality: The Big Five Factor Structure." *Journal of Personality and Social Psychology* 59, 1216–29, 1990.

Heath, A. C., C. R. Cloninger, and N. G. Martin. "Testing a Model for the Genetic Structure of Personality: A Comparison of the Personality Systems of Cloninger and Eysenck." *Journal of Personality and Social Psychology* 66, 762–75, 1994.

Hergenhahn, B. R. *An Introduction to the History of Psychology.* 2nd ed. Pacific Grove, CA: Brooks/Cole Publishing, 1992.

LeDoux, Joseph. *The Emotional Brain: The Mysterious Underpinnings of Emotional Life.* New York: Simon & Schuster, 1996.

McCrae, R. R. "Consensual Validation of Personality Traits: Evidence from Self-Reports and Ratings." *Journal of Personality and Social Psychology* 43, 293–303, 1982.

McGue, M., S. Bacon, and D. T. Lykken. "Personality Stability and Change in Early Adulthood: A Behavioral Genetic Analysis." *Developmental Psychology* 29, 96–109, 1993.

Plomin, R., and R. D. Rende. "Human Behavioral Genetics." *Annual Review of Psychology* 42, 161–90, 1991.

Plomin, R., N. L. Pedersen, P. Lichtenstein, and G. E. McClearn. "Variability and Stability in Cognitive Abilities Are Largely Genetic Later in Life." *Behavior Genetics* 24, 207–15, 1994.

Plomin, Robert. "The Role of Inheritance in Behavior." *Science* 248, 183–8, 1990.

Plomin, Robert, John C. DeFries, Gerald E. McClearn, and Michael Rutter.

Behavioral Genetics. 3rd ed. New York: W. H. Freeman and Company, 1997.

Plomin, Robert, Michael J. Owen, and Peter McGuffin. "The Genetic Basis of Complex Human Behaviors." *Science* 264, 1733–9, 1994.

Restak, Richard M. *Receptors.* New York: Bantam Books, 1994.

Rose, Richard J. "Genes and Human Behavior." *Annual Review of Psychology* 46, 625–54, 1995.

Rowe, David C. *The Limits of Family Influence: Genes, Experience, and Behavior.* New York: Guilford Press, 1994.

Schultz, Duane, and Sydney Ellen Schultz. *Theories of Personality.* 5th ed. Pacific Grove, CA: Brooks/Cole Publishing, 1994.

Sheperd, Gordon M. *Neurobiology.* 3rd ed. Oxford: Oxford University Press, 1994.

Singer, Maxine, and Paul Berg. *Genes and Genomes: A Changing Perspective.* Mill Valley, CA: University Science Books, 1991.

Tellegen, Auke, D. T. Lykken, T. J. Bouchard, K. J. Wilcox, N. L. Segal, and S. Rich. "Personality Similarity in Twins Reared Apart and Together." *Journal of Personality and Social Psychology* 54, 1031–9, 1988.

Williams, Xandria. *The Four Temperaments.* New York: St. Martin's Press, 1996.

Wingerson, Lois. *Mapping Our Genes: The Genome Project and the Future of Medicine.* New York: Plume, 1991.

Wright, Robert. *The Moral Animal.* New York: Pantheon Books, 1994.

Zuckerman, M., D. M. Kuhlman, J. Joireman, P. Teta, and M. Kraft. "A Comparison of Three Structural Models for Personality: The Big Three, the Big Five, and the Alternative Five." *The Journal of Personality and Social Psychology* 65, 757–68, 1993.

Zuckerman, Marvin. "Good and Bad Humors: Biochemical Bases of Personality and Its Disorders." *Psychological Science* 6, 325–33, 1995.

CHAPTER 1: THRILLS

Benjamin, J., L. Li, C. Patterson, B. D. Greenberg, D. L. Murphy, and D. H. Hamer. "Population and Familial Association Between the D4 Dopamine Receptor Gene and Measures of Novelty Seeking." *Nature Genetics* 12, 81–4, 1996.

Blaszczynski, A. P., A. C. Wilson, and N. McGonaghy. "Sensation Seeking and Pathological Gambling." *British Journal of Addiction* 81, 113–7, 1986.

Depue, Richard A., Monica Luciana, Paul Arbisi, Paul Collins, and Arthur Leon. "Dopamine and the Structure of Personality: Relation of Agonist-Induced Dopamine Activity to Positive Emotionality." *Journal of Personality and Social Psychology* 67, 485–98, 1994.

Donaldson, S. "Similarity in Sensation-Seeking, Sexual Satisfaction, and Contentment in Relationship in Heterosexual Couples." *Psychological Reports* 64, 405–6, 1989.

Ebstein, R. P., O. Novick, R. Umansky, B. Priel, Y. Osher, D. Blaine, E. R. Bennett, L. Nemanov, M. Katz, and R. H. Belmaker. "Dopamine D4 Receptor (D4DR) Exon III Polymorphism Associated with the Human Personality Trait of Novelty Seeking." *Nature Genetics* 12, 78–80, 1995.

Eysenck, S. B. G., and M. Zuckerman. "The Relationship Between Sensation Seeking and Eysenck's Dimensions of Personality." *British Journal of Psychology* 69, 483–7, 1978.

Ficher, I. V., M. Zuckerman, and M. Steinberg. "Sensation Seeking Congruence in Couples as a Determinant of Marital Adjustment: A Partial Replication and Extension." *Journal of Clinical Psychology* 44, 803–9, 1988.

Fink, J. S., and G. P. Smith. "Mesolimbic and Mesocortical Dopaminergic Neurons Are Necessary for Normal Exploratory Behavior in Rats." *Neuroscience Letters* 17, 61–5, 1980.

Fulker, D. W., S. B. G. Eysenck, and M. Zuckerman. "A Genetic and Environmental Analysis of Sensation Seeking." *Journal of Research in Personality* 14, 261–81, 1980.

Furnham, A. F., and M. Bunyan. "Personality and Art Preferences." *European Journal of Personality* 2, 67–74, 1988.

Giros, B., M. Jaber, S. R. Jones, R. M. Wrightman, and M. G. Caron. "Hyperlocomotion and Indifference to Cocaine and Amphetamine in Mice Lacking the Dopamine Transporter." *Nature* 379, 606–12, 1996.

Hartmann, Thom. *Attention Deficit Disorder: A Different Perception.* Penn Valley, CA: Underwood-Miller, 1993.

Menza, Matthew A., Lawrence I. Golbe, Ronald A. Cody, and Nancy E.

Forman. "Dopamine-Related Personality Traits in Parkinson's Disease." *Neurology* 43, 505–8, 1993.

McCourt, W. F., R. J. Guerrera, and H. S. Cutter. "Sensation Seeking and Novelty Seeking. Are They the Same?" *Journal of Nervous and Mental Disease* 181, 309–12, 1993.

Zhou, Qun-Yong, and Richard D. Palmiter. "Dopamine-Deficient Mice Are Severely Hypoactive, Adipsic, and Aphagic." *Cell* 83, 1197–1209, 1995.

Zuckerman, Marvin. *Behavioral Expressions and Biosocial Bases of Sensation Seeking.* Cambridge: Cambridge University Press, 1994.

———. "A Biological Theory of Sensation Seeking." In *Biological Bases of Sensation Seeking, Impulsivity, and Anxiety,* edited by Marvin Zuckerman. Hillsdale, NJ: Erlbaum, 1983.

CHAPTER 2: WORRY

Briley, M. S., S. Z. Langer, R. Raisman, D. Sechter, and E. Zarifian. "Tritiated Imipramine Binding Sites Are Decreased in Platelets from Untreated Depressed Patients." *Science* 209, 303–5, 1980.

Calkins, S. D., and N. A. Fox. "Behavioral and Physiological Antecedents of Inhibited and Uninhibited Behavior." *Child Development* 67, 523–40, 1996.

Castrogiovanni, P., I. Maremmani, A. Di Muro, A. Rotondo, and D. Marazziti. "Personality Features and Platelet ^{3}H-Imipramine Binding." *Neuropsychobiology* 25, 11–3, 1992.

Cherny, S. S., D. W. Fulker, R. N. Emde, J. Robinson, R. P. Corley, J. S. Reznick, R. Plomlin, and J. C. DeFries. "Continuity and Change in Infant Shyness from 14 to 20 Months." *Behavior Genetics* 24, 365–79, 1994.

Cytryn, Leon, and Donald McKnew. *Growing Up Sad: Childhood Depression and Its Treatment.* New York: W. W. Norton, 1996.

Flint, J., R. Corley, J. C. DeFries, D. W. Fulker, J. A. Gray, S. Miller, and A. C. Collins. "A Simple Genetic Basis for a Complex Psychological Trait in Laboratory Mice." *Science* 269, 1432–5, 1995.

Fox, Nathan A., Louis A. Schmidt, Susan D. Calkins, Kenneth H. Rubin, and Robert J. Coplan. "The Role of Frontal Activation in the Regulation and Dysregulation of Social Behavior During the Preschool Years." *Development and Psychopathology* 8, 89–102, 1996.

Griebel, Guy. "5-Hydroxytryptamine-Interacting Drugs in Animal Models of Anxiety Disorders: More Than 30 Years of Research." *Pharmacology and Therapeutics* 65, 319–95, 1995.

Handley, Sheila L. "5-Hydroxytryptamine Pathways in Anxiety and Its Treatments." *Pharmacology and Therapeutics* 66, 103–48, 1995.

Hirshfeld, Dina R., Jerrold F. Rosenbaum, Joseph Biederman, Elizabeth A. Bolduc, Stephen V. Faraone, Nancy Snidman, J. Steven Reznick, and Jerome Kagan. "Stable Behavioral Inhibition and Its Association with Anxiety Disorder." *Journal of the American Academy of Child and Adolescent Psychiatry* 31, 103–11, 1992.

Jamison, Kay Redfield. *Touched with Fire: Manic-Depressive Illness and the Artistic Temperament.* New York: Free Press, 1993.

———. *An Unquiet Mind.* New York: Knopf, 1995.

Kagan, J., J. S. Reznick, and N. Snidman. "Biological Bases of Childhood Shyness." *Science* 240, 167–71, 1988.

Kagan, Jerome. *Galen's Prophecy: Temperament in Human Nature.* New York: Basic Books, 1994.

Kendler, K. S., M. C. Neale, R. C. Kessler, A. C. Heath, and L. J. Eaves. "A Population-Based Twin Study of Major Depression in Women: The Impact of Varying Definitions of Illness." *Archives of General Psychiatry* 49, 257–66, 1992.

Kendler, K. S., M. C. Neale, R. C. Kessler, A. C. Heath, and L. J. Eaves. "Major Depression and Generalized Anxiety Disorder: Same Genes (Partly) Different Environments?" *Archives of General Psychiatry* 49, 716–22, 1992.

Kessler, R. C., A. Heath, K. S. Kendler, M. C. Neale, and L. J. Eaves. "Social Support, Depressed Mood, and Adjustment to Stress: A Genetic Epidemiologic Investigation." *Journal of Personality and Social Psychology* 62, 257–72, 1992.

Kramer, Peter D. *Listening to Prozac.* New York: Viking, 1993.

Lesch, Klaus-Peter, Dietmar Bengel, Armin Heils, Sue Z. Sabol, Bejamin D. Greenberg, Susanne Petri, Jonathan Benjamin, Clemens R. Muller, Dean H. Hamer, and Dennis L. Murphy. "Association of Anxiety-Related Traits with a Polymorphism in the Serotonin Transporter Gene Regulatory Region." *Science* 274, 1527–31, 1996.

Lykken, David T., and A. Tellegen. "Happiness Is a Stochastic Phenomenon." *Psychological Science* 7, 186–9, 1996.

Pedersen, N. L., R. Plomin, G. E. McClearn, and L. Friberg. "Neuroticism, Extraversion, and Related Traits in Adult Twins Reared Apart and Together." *Journal of Personality and Social Psychology* 55, 950–7, 1988.

Roy, Alec, Nancy L. Segal, Brandon S. Centerwall, and C. Dennis Robinette. "Suicide in Twins." *Archives of General Psychiatry* 48, 29–32, 1991.

Roy, Alec, Nancy L. Segal, and Marco Sarchiapone. "Attempted Suicide Among Living Co-Twins of Twin Suicide Victims." *American Journal of Psychiatry* 152, 1075–6, 1995.

Schmidt, Louis A., and Nathan A. Fox. "Individual Differences in Young Adults' Shyness and Sociability: Personality and Health Correlates." *Personality and Individual Differences* 19, 455–62, 1995.

Schmidt, Louis A., and Nathan A. Fox. "Patterns of Cortical Electrophysiology and Autonomic Activity in Adults' Shyness and Sociability." *Biological Psychology* 38, 183–98, 1994.

Segal, Nancy L., and Alec Roy. "Suicide Attempts in Twins Whose Co-Twins' Deaths Were Non-Suicides." *Personality and Individual Differences* 19, 937–40, 1995.

Torgersen, S. "A Twin-Study Perspective of the Comorbidity of Anxiety and Depression." In *Comorbidity of Mood and Anxiety Disorders,* edited by J. D. Maser and C. R. Cloninger. Washington, DC: American Psychiatric Press, 1990.

Yapko, Michael D. *Breaking the Patterns of Depression.* New York: Doubleday, 1997.

CHAPTER 3: ANGER

Archer, J. "The Influence of Testosterone on Human Aggression." *British Journal of Psychology* 82, 1–28, 1991.

Baer, R. A., and M. T. Nietzel. "Cognitive and Behavioral Treatment of Impulsivity in Children: A Meta-analytic Review of the Outcome of Literature." *Journal of Clinical Child Psychology* 20, 400–12, 1991.

Bock, G. R., and J. A. Goode. *Genetics of Criminal and Antisocial Behavior.* Chichester, UK: Wiley, 1996.

Bohman, M., C. R. Cloninger, S. Sigvardsson, and A. von Knorring. "Predisposition to Petty Criminality in Swedish Adoptees. I. Genetic and Environmental Heterogeneity." *Archives of General Psychiatry* 39, 1233–41, 1982.

Bohman, M., C. R. Cloninger, A. von Knorring, and S. Sigvardsson. "An Adoption Study of Somatoform Disorders: III. Cross-Fostering Analysis and Genetic Relationship to Alcoholism and Criminality." *Archives of General Psychiatry* 41, 872–8, 1984.

Brunner, H. G., M. Nelen, X. O. Breakefield, H. H. Ropers, and B. A. von Oost. "Abnormal Behavior Associated with a Point Mutation in the Structural Gene for Monoamine Oxidase A." *Science* 26, 578–80, 1993.

Brunner, H. G., M. R. Nelen, P. van Zandvoort, N. G. G. M. Abeling, A. H. van Gennip, E. C. Wolters, M. A. Kuiper, H. H. Ropers, and B. A. van Oost. "X-Linked Borderline Mental Retardation with Prominent Behavioral Disturbance: Phenotype, Genetic Localization, and Evidence for Disturbed Monoamine Metabolism." *American Journal of Human Genetics* 52, 1032–9, 1993.

Cadoret, R. J. "Psychopathology in Adopted-Away Offspring of Biological Parents with Antisocial Behavior." *Archives of General Psychiatry* 35, 1171–5, 1978.

Cadoret, R. J., and C. Cain. "Environmental and Genetic Factors in Predicting Adolescent Antisocial Behavior." *The Psychiatric Journal of the University of Ottawa* 6, 220–5, 1981.

Cadoret, R. J., C. Cain, and R. R. Crowe. "Evidence for Gene-Environment Interaction in the Development of Adolescent Antisocial Behavior." *Behavior Genetics* 13, 301–10, 1983.

Cadoret, R. J., T. W. O'Gorman, E. Troughton, and E. Heywood. "Alcoholism and Antisocial Personality: Interrelationships, Genetic, and Environmental Factors." *Archives of General Psychiatry* 42, 161–7, 1985.

Cadoret, R. J., W. R. Yates, E. Troughton, G. Woodworth, and M. A. Stewart. "Gene-Environment Interaction in the Genesis of Aggressivity and Conduct Disorders." *Archives of General Psychiatry* 52, 916–24, 1995.

Cases, O., Isabelle Seif, Joseph Grimsby, Patricia Gaspar, K. Chen, Sandrine Pournin, Ulrike Muller, Michel Aguet, Charles Babinet, Jean Chen

Shih, and Edward De Maeyer. "Aggressive Behavior and Altered Amounts of Brain Serotonin and Norepinephrine in Mice Lacking MAOA." *Science* 268, 1763–6, 1995.

Chen, Chong, Donald G. Rainnie, Robert W. Greene, and Susumu Tonegawa. "Abnormal Fear Response and Aggressive Behavior in Mutant Mice Deficient for α-Calcium-Calmodulin Kinase II." *Science* 266, 291–4, 1994.

Cloninger, C. R., M. Bohman, S. Sigvardsson, and A. von Knorring. "Predisposition to Petty Criminality in Swedish Adoptees. II. Cross-Fostering Analysis of Gene-Environment Interaction." *Archives of General Psychiatry* 39, 1242–47, 1982.

Coccaro, E. F., C. S. Bergeman, and G. E. McClearn. "Heritability of Irritable Impulsiveness: A Study of Twins Reared Together and Apart." *Psychiatry Research* 48, 229–42, 1993.

Crowe, R. R. "An Adoption Study of Antisocial Personality." *Archives of General Psychiatry* 31, 785–91, 1974.

Daly, Martin, and Margo Wilson. "Evolutionary Psychology of Male Violence." In *Male Violence,* edited by J. Archer. London: Routledge, 1994.

Daly, Martin, and Margo Wilson. "Killing the Competition: Female/Female and Male/Male Homicide." *Human Nature* 1, 81–107, 1990.

Daly, Martin, and Margo Wilson. "Evolutionary Social Psychology and Family Homicide." *Science* 242, 519–24, 1988.

Dodge, K. A. "Attributional Bias in Aggressive Children." In *Advances in Cognitive-Behavioral Research and Therapy,* edited by P. C. Kendall. Vol. 4. Orlando, FL: Academic Press, 1985.

Dodge, Kenneth A., John E. Bates, and Gregory S. Pettit. "Mechanisms in the Cycle of Violence." *Science* 250, 1678–83, 1990.

Gladue, Brian A., Michael Boechler, and Devin D. McCaul. "Hormonal Response to Competition in Human Males." *Aggressive Behavior* 15, 409–22, 1989.

Grove, W. M., E. D. Elke, L. Heston, T. J. Bouchard, Jr., N. Segal, and D. T. Lykken. "Heritability of Substance Abuse and Antisocial Behavior: A Study of Monozygotic Twins Reared Apart." *Biological Psychiatry* 27, 1294–1304, 1990.

Hen, Rene. "Mean Genes." *Neuron* 16, 1–20, 1996.

Higley, J. Dee, P. T. Mehlman, D. M. Taub, S. B. Higley, Stephen J. Suomi,

M. Linnoila, and J. H. Vickers. "Cerebrospinal Fluid Monoamine and Adrenal Correlates of Aggression in Free-Ranging Rhesus Monkeys." *Archives of General Psychiatry* 49, 436–41, 1992.

Kazdin, Alan E. *Conduct Disorders in Childhood and Adolescence.* 2nd ed. Thousand Oaks, CA: Sage Publications, 1995.

Lyons, M. J., W. R. True, S. A. Eisen, J. Goldberg, J. M. Meyer, S. V. Faraone, L. J. Eaves, and M. T. Tsuang. "Differential Heritability of Adult and Juvenile Antisocial Traits." *Archives of General Psychiatry* 52, 906–15, 1995.

McCaul, Kevin D., Brian A. Gladue, and Margaret Joppe. "Winning, Losing, Mood, and Testosterone." *Hormones and Behavior* 26, 486–504, 1992.

Mednick, S. A., W. F. Gabrielli, and B. Hutchings. "Genetic Factors in Criminal Behavior: Evidence from an Adoption Cohort." *Science* 224, 891–3, 1984.

Moran, Timothy H., Roger H. Reeves, Derek Rogers, and Elizabeth Fisher. "Ain't Misbehavin'—It's Genetic!" *Nature Genetics* 12, 115–6, 1996.

Nelson, R. J., G. E. Demas, P. L. Huang, M. C. Fishman, V. L. Dawson, T. M. Dawson, and S. H. Snyder. "Behavioural Abnormalities in Male Mice Lacking Neuronal Nitric Oxide Synthase." *Nature* 378, 383–6, 1995.

Raleigh, M. J., and M. T. McGuire. "Bidirectional Relationships Between Typtophan and Social Behavior in Vervet Monkeys." *Advances in Experimental Medicine and Biology* 294, 289–98, 1991.

Raleigh, M. J., M. T. McGuire, G. L. Brammer, D. B. Pollack and A. Yuwiler. "Serotonergic Mechanisms Promote Dominance Acquisition in Adult Male Vervet Monkeys." *Brain Research* 559, 181–90, 1991.

Reiss, David, Mavis Hetherington, Robert Plomin, George W. Howe, Samuel J. Simmens, Sandra H. Henderson, Thomas J. O'Connor, Danielle A. Bussell, Edward R. Anderson, and Tracy Law. "Differential Parenting and Psychopathology in Adolescence." *Archives of General Psychiatry* 52, 925–36, 1995.

Saudou, F., D. A. Amara, A. Dierich, M. Lemur, S. Ramboz, L. Segu, M. C. Buhot, and R. Hen. "Enhanced Aggressive Behavior in Mice Lacking 5-HT$_{1B}$ Receptor." *Science* 265, 1875–8, 1994.

CHAPTER 4: ADDICTION

Alcoholics Anonymous. 3rd ed. New York: Alcoholics Anonymous World Services, Inc., 1976.

Bartecchi, Carl E., Thomas D. MacKenzie, and Robert W. Schrier. "The Global Tobacco Epidemic." *Scientific American* 272, 44–51, 1995.

Blum, K., and E. Noble. "Allelic Association of Human Dopamine D_2 Receptor Gene in Alcoholism." *Journal of the American Medical Association* 263, 2055–60, 1990.

Buck, Kari Johnson. "Molecular Genetic Analysis of the Role of GABAergic Systems in the Behavioral and Cellular Actions of Alcohol." *Behavior Genetics* 26, 313–23, 1996.

Cloninger, C. R. "Neurogenetic Adaptive Mechanisms in Alcoholism." *Science* 236, 410–6, 1987.

Cloninger, C. R., M. Bohman, and S. Sigvardsson. "Inheritance of Alcohol Abuse: Cross-Fostering Analysis of Adopted Men." *Archives of General Psychiatry* 38, 861–8, 1981.

Cloninger, C. Robert, and Henri Begleiter, eds. *Genetics and Biology of Alcoholism.* Plainview, NY: Cold Spring Harbor Laboratory Press, 1990.

Crabbe, John C., John K. Belknap, and Kari J. Buck. "Genetic Animal Models of Alcohol and Drug Abuse." Science 264, 1715–23, 1994.

Crabbe, John C., Tamara J. Phillips, Daniel J. Feller, Rene Hen, Charlotte D. Wenger, Christina N. Lessov, and Gwen L. Schafer. "Elevated Alcohol Consumption in Null Mutant Mice Lacking 5-HT_{1B} Serotonin Receptors." *Nature Genetics* 14, 98–101, 1996.

Crabbe, John C., Jr., and R. Adron Harris, eds. *The Genetic Basis of Alcohol and Drug Actions.* New York: Plenum Press, 1991.

Fowler, J. S., N. D. Volkow, G.-J. Wang, N. Pappas, J. Logan, R. MacGregor, D. Alexoff, C. Shea, D. Schyler, A. P. Wolf, D. Warner, I. Zezulkova, and R. Cilento. "Inhibition of Monoamine Oxidase B in the Brains of Smokers." *Nature* 379, 733–6, 1996.

Gejman, Pablo V., Anca Ram, Joel Gelernter, Eitan Friedman, Qiuhe Cao, David Pickar, Kenneth Blum, Ernest P. Noble, Henry R. Kranzler, Stephanie O'Malley, Dean H. Hamer, Flanagan Whitsitt, Peter Rao, Lynn E. DeLisi, Matti Vikkunen, Markku Linnoila, David Goldman, and Elliot S. Gershon. "No Structural Mutation in the Dopamine D_2

Receptor Gene in Alcoholism or Schizophrenia: Analysis Using Denaturing Gradient Gel Electrophoresis." *Journal of the American Medical Association* 271, 204–8, 1994.

Gelernter, J., D. Goldman, and N. Risch. "The Al Allele at the D₂ Dopamine Receptor Gene and Alcoholism: A Reappraisal." *Journal of the American Medical Association* 269, 1673–7, 1993.

Grove, W. M., E. D. Eckert, L. Heston, T. J. Bouchard, Jr., N. Segal, and D. T. Lykken. "Heritability of Substance Abuse and Antisocial Behavior: A Study of Monozygotic Twins Reared Apart." *Biological Psychiatry* 27, 1293–1304, 1990.

Heath, A. C., and N. Martin. "Genetic Models for the Natural History of Smoking: Evidence for a Genetic Influence on Smoking Persistence." *Addictive Behaviors* 18, 19–34, 1993.

Heath, A. C., Pamela A. F. Madden, Wendy S. Slutske, and Nicholas G. Martin. "Personality and the Inheritance of Smoking Behavior: A Genetic Perspective." *Behavior Genetics* 25, 103–17, 1995.

Heath, A. C., and P. G. Madden. "Genetic Influences on Smoking Behavior." In *Behavior Genetic Approaches in Behavioral Medicine,* edited by J. R. Turner, L. R. Cardon, and J. K. Hewitt. New York: Plenum, 1995.

Kassel, Jon D., Saul Shiffman, Maryann Gnys, Jean Paty, and Monica Zettler-Segal. "Psychosocial and Personality Differences in Chippers and Regular Smokers." *Addictive Behaviors* 19, 565–75, 1994.

Kendler, K. S., A. C. Heath, M. C. Neale, R. C. Kessler, and L. J. Eaves. "A Population-Based Twin Study of Alcoholism in Women." *Journal of American Medicine* 268, 1877–82, 1992.

Overstreet, David H. "Differential Effects of Nicotine in Inbred and Selectively Bred Rodents." *Behavior Genetics* 25, 179–96, 1995.

Patton, David, Gordon E. Barnes, and Robert P. Murray. "Personality Characteristics of Smokers and Ex-Smokers." *Personality and Individual Differences* 15, 653–64, 1993.

Pomerleau, C. S., O. F. Pomerleau, K. A. Flessland, and S. M. Basson. "Relationship of Tridimensional Personality Questionnaire Scores and Smoking Variables in Female and Male Smokers." *Journal of Substance Abuse* 4, 143–54, 1992.

Pomerleau, Ovide F. "Individual Differences in Sensitivity to Nicotine: Implications for Genetic Research on Nicotine Dependence." *Behavior Genetics* 25, 161–77, 1995.

Pontieri, Francesco E., Gianluigi Tanda, Francesco Orzi, and Gaetano Di Chiara. "Effects of Nicotine on the Nucleus Accumbens and Similarity to Those of Addictive Drugs." *Nature* 382, 255–7, 1996.

Reto, R., A. D. Lopez, J. Boreham, M. Thun, and C. Heath. "Mortality from Tobacco in Developed Countries: Indirect Estimation from National Vital Statistics." *Lancet* 339, 1268–78, 1992.

Rowe, D. C., and M. R. Linver. "Smoking and Addictive Behaviors: Epidemiological, Individual, and Family Factors." In *Behavior Genetic Approaches in Behavioral Medicine,* edited by J. R. Turner, L. R. Cardon, and J. K. Hewitt. New York: Plenum, 1995.

Self, D. W., and E. J. Nestler. "Molecular Mechanisms of Drug Reinforcement and Addiction." *Annual Review of Neuroscience* 18, 463–95, 1995.

Sun, Grace Y., P. Kevin Rudeen, W. Gibson Wood, Yau Huei Wei, and Albert Y. Sun, eds. *Molecular Mechanisms of Alcohol: Neurobiology and Metabolism.* Clifton, NJ: Humana Press, 1989.

Swan, G. E., D. Carmelli, R. H. Rosenman, R. H. Fabsitz, and J. C. Christian. "Smoking and Alcohol Consumption in Adult Male Twins: Genetic Heritability and Shared Environmental Influences." *Journal of Substance Abuse* 2, 39–50, 1990.

Uhl, G., K. Blum, E. P. Nobel, and S. Smith. "Substance Abuse Vulnerability and D_2 Dopamine Receptor Gene and Severe Alcoholism." *Trends in Neuroscience* 16, 83–8, 1993.

Vaillant, George E. *The Natural History of Alcoholism: Causes, Patterns, and Paths to Recovery.* Cambridge, MA: Harvard University Press, 1983.

Zuckerman, Marvin. "Sensation Seeking: The Initial Motive for Drug Abuse." In *Etiological Aspects of Alcohol and Drug Abuse,* edited by E. H. Gottheil, K. A. Druley, T. E. Skoloda, and H. H. Waxman. Springfield, IL: Charles C. Thomas, 1983.

CHAPTER 5: SEX

Abramson, Paul R., and Steven D. Pinkerton, eds. *Sexual Nature, Sexual Culture.* Chicago: Chicago University Press, 1995.

Bailey, J. M., and R. C. Pillard. "A Genetic Study of Male Sexual Orientation." *Archives of General Psychiatry* 48, 1089–96, 1991.

Bailey, J. M., R. C. Pillard, M. C. Neale, and Y. Agyei. "Heritable Factors Influence Sexual Orientation in Women." *Archives of General Psychiatry* 50, 217–23, 1993.

Bogaert, Anthony F., and William A. Fisher. "Predictors of University Men's Number of Sexual Partners." *The Journal of Sex Research* 32, 119–30, 1995.

Bullough, Vern L. *Science in the Bedroom: A History of Sexual Research.* New York: Basic Books, 1994.

Burr, Chandler. *A Separate Creation: The Search for the Biological Origins of Sexual Orientation.* New York: Hyperion, 1996.

Buss, D. M. "Sex Differences in Human Mate Preferences: Evolutionary Hypotheses Tested in 37 Cultures." *Behavioral and Brain Sciences* 12, 1–49, 1989.

Buss, David M. *The Evolution of Desire: Strategies of Human Mating.* New York: Basic Books, 1994.

Diamond, M., and H. K. Sigmundson. "Sex Assignment at Birth: Long-Term Review and Clinical Implications." *Archives of Pediatric and Adolescent Medicine* 151, 298–304, 1997.

Fisher, Helen E. *Anatomy of Love: The Natural History of Monogamy, Adultery, and Divorce.* New York: W. W. Norton, 1992.

Friedman, Richard C., and Jennifer I. Downey. "Homosexuality." *The New England Journal of Medicine* 331, 923–30, 1994.

Hall, J. "Courtship Among Males Due to a Male-Sterile Mutation in *Drosophila Melanogaster.*" *Behavioral Genetics* 8, 125–41, 1978.

Hall, Jeffrey C. "The Mating of a Fly." *Science* 264, 1702–14, 1994.

Hamer, D. H., S. Hu, V. L. Magnuson, N. Hu, and A. M. L. Pattatucci. "A Linkage Between DNA Markers on the X Chromosome and Male Sexual Orientation." *Science* 261, 321–7, 1993.

Hamer, Dean, and Peter Copeland. *The Science of Desire: The Search for the Gay Gene and the Biology of Behavior.* New York: Simon & Schuster, 1994.

Hu, S., A. M. L. Patatucci, C. Patterson, L. Li, D. W. Fulker, S. S. Cherny, L. Kruglyak, and D. Hamer. "Linkage Between Sexual Orientation and Chromosome Xq28 in Males But Not in Females." *Nature Genetics* 11, 248–56, 1995.

Kafka, Martin P. "Sertraline Pharmacotherapy for Paraphilias and Paraphilia-Related Disorders: An Open Trial." *Annals of Clinical Psychiatry* 6, 189–95, 1994.

———. "Successful Treatment of Paraphilic Coercive Disorder (a Rapist) with Fluoxetine Hydrochloride." *British Journal of Psychiatry* 158, 844–7, 1991.

———. "Successful Antidepressant Treatment of Nonparaphilic Sexual Addictions and Paraphilias in Men." *Journal of Clinical Psychiatry* 52, 60–5, 1991.

Kallman, F. J. "Twin and Sibship Study of Overt Male Homosexuality." *American Journal of Human Genetics* 4, 136–46, 1952.

LeVay, S. "A Difference in Hypothalamic Structure Between Heterosexual and Homosexual Men." *Science* 253, 1034–7, 1991.

LeVay, S., and D. H. Hamer. "Evidence for a Biological Influence in Male Homosexuality." *Scientific American* 270, 44–9, 1994.

LeVay, Simon. *The Sexual Brain.* Cambridge, MA: Massachusetts Institute of Technology Press, 1993.

Mealey, Linda, and Nancy L. Segal. "Heritable and Environmental Variables Affect Reproduction-Related Behaviors, But Not Ultimate Reproductive Success." *Personality and Individual Differences* 14, 783–94, 1993.

Pattatucci, Angela M. L., and Dean H. Hamer. "Development and Familiarity of Sexual Orientation in Females." *Behavior Genetics* 25, 407–19, 1995.

Pool, Robert. *Eve's Rib: Searching for the Biological Roots of Sex Differences.* New York: Crown, 1994.

Ridley, Matt. *The Red Queen: Sex and the Evolution of Human Nature.* New York: Macmillan, 1993.

Small, Meredith F. *What's Love Got to Do with It? The Evolution of Human Mating.* New York: Anchor Books, 1995.

Stevens, George, and Robert Bellig, eds. *The Evolution of Sex.* New York: Harper and Row, 1988.

Walsh, Anthony. *The Science of Love: Understanding Love and Its Effects on Mind and Body.* Buffalo, NY: Prometheus Books, 1991.

Zhang, S., and W. Odenwald. "Misexpression of the White (W) Gene Triggers Male-Male Courtship in *Drosophila*." *Proceedings of the National Academy of Sciences, USA* 92, 5525–9, 1995.

CHAPTER 6: THINKING

Bailey, C. H., D. Bartsch, and E. R. Kandel. "Toward a Molecular Definition of Long-Term Memory Storage." *Proceedings of the National Academy of Sciences, USA* 93, 12445–52, 1996.

Bishop, D. V. M., T. North, and C. Donlan. "Genetic Basis of Specific Language Impairment: Evidence from a Twin Study." *Developmental Medicine and Child Neurology* 37, 56–71, 1995.

Bouchard, T. J., Jr., and M. McGue. "Familial Studies of Intelligence: A Review." *Science* 212, 1055–9, 1981.

Brooks, A., D. W. Fulker, and J. C. DeFries. "Reading Performance and General Cognitive Ability: A Multivariate Genetic Analysis of Twin Data." *Personality and Individual Differences* 11, 141–6, 1990.

Capron, C., and M. Duyme. "Assessment of Effects of Socio-Economic Status on IQ in a Full Cross-Fostering Study." *Nature* 340, 552–3, 1989.

Cardon, L. R., S. D. Smith, D. W. Fulker, W. J. Kimberling, B. F. Pennington, and J. C. DeFries. "Quantitative Trait Locus for Reading Disability on Chromosome 6." *Science* 266, 276–9, 1994.

DeFries, J. C., and M. Alarcon. "Genetics of Specific Reading Disability." *Mental Retardation and Developmental Disabilities Research Reviews* 2, 39–47, 1996.

DeFries, J. C., D. W. Fulker, and M. C. LaBuda. "Evidence for a Genetic Aetiology in Reading Disability of Twins." *Nature* 329, 537–9, 1987.

Fulker, D. W., J. C. DeFries, and R. Plomin. "Genetic Influence on General Mental Ability Increases Between Infancy and Middle Childhood." *Nature* 336, 767–9, 1988.

Gilger, Jeffrey W. "Behavioral Genetics: Concepts for Research and Practice in Language Development and Disorders." *Journal of Speech and Hearing Research* 38, 1126–42, 1995.

Herrnstein, Richard J., and Charles Murray. *The Bell Curve: Intelligence and Class Structure in American Life.* New York: The Free Press, 1994.

McCartney, K., M. J. Harris, and F. Bernieri. "Growing Up and Growing Apart: A Developmental Meta-Analysis of Twin Studies." *Psychology Bulletin* 107, 226–37, 1990.

McGue, M., T. J. Bouchard, Jr., W. G. Iacono, and D. T. Lykken. "Behavioral Genetics of Cognitive Ability: A Life-Span Perspective." In *Na-*

ture, Nurture, and Psychology, edited by R. Plomin and G. E. Mc-Clearn. Washington, DC: American Psychological Association, 1993.

Meryash, D. L., C. E. Cronk, B. Sachs, and P. S. Gerald. "An Anthropometric Study of Males with the Fragile-X Syndrome." *American Journal of Medical Genetics* 17, 159–74, 1984.

Pinker, Steven. *The Language Instinct: How the Mind Creates Language.* New York: Harper Perennial, 1994.

Plomin, Robert, Gerald E. McClearn, Deborah L. Smith, Sylvia Vignetti, Michael J. Chorney, Karen Chorney, Charles P. Venditti, Steven Kasarda, Lee A. Thompson, Douglas K. Detterman, Johanna Daniels, Michael Owen, and Peter McGuffin. "DNA Markers Associated with High Versus Low IQ: The IQ Quantitative Trait Loci (QTL) Project." *Behavior Genetics* 24, 107–18, 1994.

Scarr, S., and R. A. Weinberg. "IQ Test Performance of Black Children Adopted by White Families." *American Psychologist* 31, 726–39, 1976.

Scriver, C. R., R. C. Eisensmith, S. L. C. Woo, and S. Kaufman. "The Hyperphenylalaninemias of Man and Mouse." *Annual Review of Genetics* 28, 141–65, 1994.

Silva, A. J., R. Paylor, J. M. Wehner, and S. Tonegawa. "Impaired Spatial Learning in α-Calcium-Calmodulin Kinase Mutant Mice." *Science* 257, 206–11, 1992.

Steele, Claude M., and Joshua Aronson. "Stereotype Threat and the Intellectual Test Performance of African Americans." *Journal of Personality and Social Psychology* 69, 7997–811, 1995.

Tully, T. "Regulation of Gene Expression and Its Role in Long-Term Memory and Synaptic Plasticity." *Proceedings of the National Academy of Sciences, USA* 94, 4239–41, 1997.

Wehner, Jeanne M., Barbara J. Bowers, and Richard Paylor. "The Use of Null Mutant Mice to Study Complex Learning and Memory Processes." *Behavior Genetics* 26, 301–12, 1996.

Whalstrom, J. "Gene Map of Mental Retardation." *Journal of Mental Deficiency Research* 34, 11–27, 1990.

Wilson, R. S. "The Louisville Twin Study: Developmental Synchronies in Behavior." *Child Development* 54, 298–316, 1983.

Yu, S., M. Pritchard, E. Kremer, M. Lynch, J. Nancarrow, E. Baker, K. Holman, J. C. Mulley, S. T. Warren, D. Schlessinger, G. R. Sutherland,

and R. I. Richards. "Fragile X Genotype Characterized by an Unstable Region of DNA." *Science* 252, 1179–81, 1991.

CHAPTER 7: HUNGER

Allison, D. B., J. Kaprio, M. Korkeila, M. Koskenvuo, M. C. Neale, and K. Hayakawa. "The Heritability of Body Mass Among an International Sample of Monozygotic Twins Reared Apart." *International Journal of Obesity Related Metabolic Disorders* 20, 501–6, 1996.

Allison, D. B., S. Heshka, M. C. Neale, D. T. Lykken, and S. B. Heymsfield. "A Genetic Analysis of Relative Weight Among 4,020 Twin Pairs, with an Emphasis on Sex Effects." *Health Psychology* 13, 362–5, 1994.

Arner, Peter. "The β_3-Adrenergic-Receptor—A Cause and Cure of Obesity?" *The New England Journal of Medicine* 333, 382–3, 1995.

Brewerton, T. D., L. D. Hand, and E. R. Bishop, Jr. "The Tridimensional Personality Questionnaire in Eating Disorder Patients." *International Journal of Eating Disorders* 14, 213–8, 1993.

Cardon, L. R. "Height, Weight, and Obesity." In *Nature and Nurture During Middle Childhood*, edited by J. C. DeFries, R. Plomin, and D. W. Fulker. Cambridge, MA: Blackwell, 1994.

———. "Genetic Influences on Body Mass Index in Early Childhood." In *Behavior Genetic Approaches in Behavioral Medicine*, edited by J. R. Turner, L. R. Cardon, and J. K. Hewitt. New York: Plenum, 1995.

Carey, D. G., T. V. Nguyen, L. V. Campbell, D. J. Chisolm, and P. Kelly. "Genetic Influences on Central Abdominal Fat: A Twin Study." *International Journal of Obesity Related Metabolic Disorders* 20, 722–6, 1996.

Chua, S. C., Jr., W. K. Chung, X. S. Wu-Peng, Y. Zhang, S.-M. Liu, L. Tartaglia, and R. L. Leibel. "Phenotypes of Mouse *Diabetes* and Rat *Fatty* Due to Mutations in the OB (Leptin) Receptor." *Science* 271, 994–6, 1996.

Clement, Karine, Christian Vaisse, Brian St. J. Manning, Arnaud Basdevant, Bernard Guy-Grand, Juan Ruiz, Kristi D. Silver, Alan R. Shuldiner, Philippe Froguel, and A. Donny Strosberg. "Genetic Variation in the β_3-Adrenergic-Receptor and an Increased Capacity to Gain Weight in

Patients with Morbid Obesity." *The New England Journal of Medicine* 333, 352–4, 1995.

Fabsitz, R. R., P. Shilinsky, and D. Carmelli. "Genetic Influences on Adult Weight Gain and Maximum Body Mass Index in Male Twins." *American Journal of Epidemiology* 140, 711–20, 1994.

Fleury, Christophe, Maria Neverova, Sheila Collins, Serge Raimbault, Odette Champigny, Corinne Levi-Meyrueis, Frederic Bouillaud, Michael F. Seldin, Richard S. Surwit, Daniel Ricquier, and Craig H. Warden. "Uncoupling Protein-2: A Novel Gene Linked to Obesity and Hyperinsulinemia." *Nature Genetics* 15, 269–72, 1997.

Gura, Trisha. "Obesity Sheds Its Secrets." *Science* 275, 751–3, 1997.

Halaas, J. L., K. S. Gajiwala, M. Maffei, S. L. Cohen, B. T. Chait, D. Rabinowitz, R. L. Lallone, S. K. Burley, and J. M. Friedman. "Weight-Reducing Effects of the Plasma Protein Encoded by the *Obese* Gene." *Science* 269, 543–6, 1995.

Hamilton, Bradford S., Diana Paglia, Anita Y. M. Kwan, and Mervyn Deitel. "Increased Obese mRNA Expression in Omental Fat Cells from Massively Obese Humans." *Nature Medicine* 1, 953–6, 1995.

Kendler, K. S., C. McLean, M. Neale, R. Kessler, A. Heath, and L. J. Eaves. "The Genetic Epidemiology of Bulimia Nervosa." *American Journal of Psychiatry* 148, 1627–37, 1991.

Lee, Gwo-Hwa, R. Proenca, J. M. Montez, K. M. Carroll, J. G. Darvishzadeh, J. I. Lee, and J. M. Friedman. "Abnormal Splicing of the Leptin Receptor in Diabetic Mice." *Nature* 379, 632–5, 1996.

Lonnqvist, Fredrik, Peter Arner, Louise Nordfors, and Martin Schalling. "Overexpression of the Obese (ob) Gene in Adipose Tissue of Human Obese Subjects." *Nature Medicine* 1, 950–2, 1995.

Maffei, M., J. Halaas, E. Ravussin, R. E. Pratley, G. H. Lee, Y. Zhang, H. Fei, S. Kim, R. Lallone, S. Ranganathan, P. A. Kern, and J. M. Friedman. "Leptin Levels in Human and Rodent: Measurement of Plasma Leptin and ob RNA in Obese and Weight-Reduced Subjects." *Nature Medicine* 1, 1155–61, 1995.

Meyer, J. M. "Genetic Studies of Obesity Across the Life Span." In *Behavior Genetic Approaches to Behavioral Medicine,* edited by J. R. Turner, L. R. Cardon, and J. K. Hewitt. New York: Plenum, 1995.

Naggert, Jurgen K., Lloyd D. Fricker, Oleg Varlamov, Patsy M. Nishina, Yves Rouille, Donald F. Steiner, Raymond J. Carroll, Beverly J.

Paigen, and Edward H. Leiter. "Hyperproinsulinaemia in Obese Fat/ Fat Mice Associated with a Carboxypeptidase E Mutation Which Reduces Enzyme Activity." *Nature Genetics* 10, 135–42, 1995.

Pelleymounter, Mary Ann, Mary Jane Cullen, Mary Beth Baker, Randy Hecht, Dwight Winters, Thomas Boone, and Frank Collins. "Effects of the Obese Gene Product on Body Weight Regulation in ob/ob Mice." *Science* 269, 540–9, 1995.

Price, R. A., R. Ness, and P. Laskarzewski. "Common Major Gene Inheritance of Extreme Overweight." *Human Biology* 62, 747–65, 1990.

Ravussin, E., M. E. Valencia, J. Esparza, P. H. Bennett, and L. O. Schulz. "Effects of a Traditional Lifestyle on Obesity in Pima Indians." *Diabetes Care* 17, 1067–74, 1994.

Rice, T., L. Perusse, C. Bouchard, and D. C. Rao. "Familial Clustering of Abdominal Visceral Fat and Total Fat Mass: The Quebec Family Study." *Obesity Research* 4, 253–61, 1996.

Sorensen, T. I. "The Genetics of Obesity." *Metabolism* 44, 4–6, 1995.

Spelt, J. R., and J. M. Meyer. "Genetics and Eating Disorders." In *Behavior Genetic Approaches in Behavioral Medicine,* edited by J. R. Turner, L. R. Cardon, and J. K. Hewitt. New York: Plenum, 1995.

Stephens, Thomas W., Margret Basinski, Pamela K. Bristow, Juliana M. Bue-Valeskey, Stanley G. Burgett, Libbey Craft, John Hale, James Hoffmann, Hansen M. Hsiung, Aidas Kriauciunas, Warren MacKellar, Paul Rosteck, Jr., Brigitte Schoner, Dennis Smith, Frank C. Tinsley, Xing-Yue Zhang, and Mark Heiman. "The Role of Neuropeptide Y in the Antiobesity of the Obese Gene Product." *Nature* 377, 530–2, 1995.

Stunkard, A. J., J. R. Harris, N. L. Perersen, and G. E. McClearn. "The Body-Mass Index of Twins Who Have Been Reared Apart." *The New England Journal of Medicine* 322, 1483–7, 1990.

Stunkard, A. J., T. T. Foch, and Z. Hrubec. "A Twin Study of Human Obesity." *Journal of the American Medical Association* 256, 561–4, 1986.

Tartaglia, Louis A., Marlene Dembaski, Xun Weng, Nanhua Deng, Janice Culpepper, Rene Devos, Grayson J. Richards, L. Arthur Campfield, Frederich T. Clark, Jim Deeds, Craig Muir, Sean Sanker, Ann Moriarty, Karen J. Moore, John S. Smutko, Gail G. Mays, Elizabeth A. Woolf, Cheryl A. Monroe, and Robert I. Tepper. "Identification and

Expression Cloning of a Leptin Receptor, OB-R." *Cell* 83, 1263–71, 1995.

Walston, Jeremy, Kristi Silver, Clifton Bogardus, William Knowler, Francesco S. Celi, Sharon Austin, Brian Maning, A. Donny Strosberg, Michael P. Stearn, Nina Raben, John D. Sorkin, Jesse Roth, and Alan Shuldiner. "Time of Onset of Non-Insulin-Dependent Diabetes Mellitus and Genetic Variation in the β_3-Adrenergic-Receptor Gene." *The New England Journal of Medicine* 333, 343–7, 1995.

Widen, Elisabeth, Markku Lehto, Timo Kanninen, Jeremy Walston, Alan R. Shuldiner, and Leif C. Groop. "Association of a Polymorphism in the β_3-Adrenergic-Receptor Gene with Features of the Insulin Resistance Syndrome in Finns." *The New England Journal of Medicine* 333, 348–51, 1995.

Wurtman, Judith, Richard Wurtman, Sharon Reynolds, Rita Tsay, Beverly Chew. "Fenfluramine Suppresses Snack Intake Among Carbohydrate Cravers But Not Among Noncarbohydrate Cravers." *International Journal of Eating Disorders* 6, 687–99, 1987.

Wurtman, Richard J., and Judith J. Wurtman. "Carbohydrates and Depression." *Scientific American* 260, 68–75, 1989.

Zhang, Yiying, Ricardo Proenca, Margherita Maffei, Marisa Barone, Lori Leopold, and Jeffry M. Friedman. "Positional Cloning of the Mouse Obese Gene and Its Human Homologue." *Nature* 372, 425–32, 1994.

CHAPTER 8: AGING

Abbott, M. H., E. A. Murphy, D. R. Bolling, and H. Abbey. "The Familial Component in Longevity, A Study of Offspring of Nonagenarians. II. Preliminary Analysis of the Completed Study." *Johns Hopkins Medical Journal* 134, 1–16, 1974.

Chiu, C. P., and C. B. Harley. "Replicative Senescence and Cell Immortality: The Role of Telomeres and Telomerase." *Proceedings of the Society for Experimental Biology and Medicine* 214, 99–106, 1997.

Corder, E. H., A. M. Saunders, N. J. Risch, W. J. Strittmatter, D. E. Shmechel, P. C. Gaskell, Jr., J. B. Rimmler, P. A. Locke, P. M. Conneally, K. E. Schmader, et al. "Protective Effect of Apolipoprotein E

Type 2 Allele for Late Onset Alzheimer Disease." *Nature Genetics* 7, 180–4, 1994.

Corder, E. H., A. M. Saunders, W. J. Strittmatter, D. E. Shmechel, P. C. Gaskell, G. W. Small, A. D. Roses, J. L. Haines, and M. A. Pericak-Vance. "Gene Dose of Apolipoprotein E Type 4 Allele and the Risk of Alzheimer's Disease in Late Onset Families." *Science* 261, 921–3, 1993.

Corral-Debrinski, M., T. Horton, M. T. Lott, J. M. Shoffner, M. F. Beal, and D. C. Wallace. "Mitochondrial DNA Deletions in Human Brain: Regional Variability and Increase with Advanced Age." *Nature Genetics* 2, 324–9, 1992.

Costa, P. T., Jr., and R. R. McCrae. "Still Stable After All These Years: Personality As a Key to Some Issues in Adulthood and Old Age." In *Life Span Development and Behavior,* edited by P. B. Baltes and O. G. Brim, Jr. Vol. III. New York: Academic Press, 1980.

Costa, Paul T., Jr., Robert R. McCrae, and Arthur H. Norris. "Personal Adjustment to Aging: Longitudinal Prediction from Neuroticism and Extraversion." *Journal of Gerontology* 1, 78–85, 1981.

Harding, A. E. "Growing Old: The Most Common Mitochondrial Disease of All?" *Nature Genetics* 2, 251–2, 1992.

Herskind, A. M., M. McGue, N. V. Holm, T. I. A. Sorensen, B. Harvald, and J. W. Vaupel. "The Heritability of Human Longevity: A Population-Based Study of 2,872 Danish Twin Pairs Born 1870–1900." *Human Genetics* 97, 319–23, 1996.

Klein, David C., Robert Y. Moore, and Steven M. Reppert. *Suprachiasmatic Nucleus: The Mind's Clock.* Oxford: Oxford University Press, 1991.

Lakowski, B., and S. Hekimi. "Determination of Life-Span in Caenorhabditis Elegans by Four Clock Genes." *Science* 272, 1010–3, 1996.

Lithgow, G. J. "Invertebrate Gerontology: The Age Mutations of Caenorhabditis Elegans." *Bioessays* 18, 809–15, 1996.

Martin, George M., Steven N. Austad, and Thomas E. Johnson. "Genetic Analysis of Aging: Role of Oxidative Damage and Environmental Stresses." *Nature Genetics* 13, 25–34, 1996.

McGue, M., J. W. Vaupel, N. Holm, and B. Harvald. "Longevity Is Moderately Heritable in a Sample of Danish Twins Born 1870–1880." *Journals of Gerontology* 48, B237–44, 1993.

Nelson, James Lindemann, and Hilde Lindemann Nelson. *Alzheimer's: Answers to Hard Questions for Families.* New York: Doubleday, 1996.

Nusbaum, T. J., and M. R. Rose. "Aging in Drosophilia." *Comparative Biochemistry and Physiology. A. Physiology* 109, 33–8, 1994.

Pericak-Vance, M. A., J. L. Bebout, P. C. Gaskell, Jr., L. H. Yamaoka, W.-Y. Hung, M. J. Alberts, A. P. Walker, R. J. Bartlett, C. A. Haynes, K. A. Welsh, N. L. Earl, A. Heyman, C. M. Clark, and A. D. Roses. "Linkage in Familial Alzheimer Disease: Evidence for Chromosome 19 Linkage." *American Journal of Human Genetics* 48, 1034–50, 1991.

Plomin, R., and G. E. McClearn. "Human Behavioral Genetics of Aging." In *Handbook of the Psychology of Aging,* edited by J. E. Birren and K. W. Schaie. New York: Academic Press, 1990.

Plomin, R., P. Lichtenstein, N. L. Pedersen, G. E. McClearn, and J. R. Nesselroade. "Genetic Influence on Life Events During the Last Half of the Life Span." *Psychology and Aging* 5, 25–30, 1990.

Selkoe, Dennis J. "Alzheimer's Disease: Genotypes, Phenotypes, and Treatments." *Science* 275, 630–2, 1997.

Schachter, F., D. Cohen, and T. Kirkwood. "Prospects for the Genetics of Human Longevity." *Human Genetics* 91, 519–26, 1993.

St. George-Hyslop, P., J. Haines, E. Rogaev, M. Mortilla, G. Vaula, M. Pericak-Vance, J.-F. Fonsin, M. Montesi, A. Bruni, S. Sorbi, et al. "Genetic Evidence for a Novel Familial Alzheimer's Disease Locus on Chromosome 14." *Nature Genetics* 2, 330–4, 1992.

Strittmatter, Warren J., Ann M. Saunders, Donald Schmechel, Margaret Pericak-Vance, Jan Enghild, Guy S. Salvesen, and Allen D. Roses. "Apolipoprotein E: High-Avidity Binding to β-Amyloid and Increased Frequency of Type 4 Allele in Late-Onset Familial Alzheimer Disease." *Proceedings of the National Academy of Sciences, USA* 90, 1977–81, 1993.

Watt, P. M., I. D. Hickson, R. H. Borts, and E. J. Louis. "SGS1, A Homologue of the Bloom's and Werner's Syndrome Genes, Is Required for Maintenance of Genome Stability in Saccharomyces Cerevisiae." *Genetics* 144, 935–45, 1996.

Yu, Chang-En, Junko Oshima, Ying-Hui Fu, Ellen M. Wijsman, Fuki Hisama, Reid Alisch, Shellie Matthews, Jun Nakura, Tetsuro Miki, Samir Ouasis, George M. Martin, John Mulligan, Gerard D. Schellenberg. "Positional Cloning of the Werner's Syndrome Gene." *Science* 272, 258–62, 1996.

CONCLUSION: ENGINEERING TEMPERAMENT

Alper, Joseph S., and Jonathan Beckwith. "Genetic Fatalism and Social Policy: The Implications of Behavior Genetics Research." *Yale Journal of Biology and Medicine* 66, 511–24, 1993.

Blomer, U., L. Naldimi, I. M. Verma, D. Trono, and F. H. Gage. "Applications of Gene Therapy to the CNS." *Human Molecular Genetics* 5, 1397–1404, 1996.

Capecchi, M. R. "Targeted Gene Replacement." *Scientific American* 270, 52–9, 1994.

Cook-Deegan, Robert. *The Gene Wars: Science, Politics, and the Human Genome.* New York: W. W. Norton, 1994.

Geller, Lisa N., Joseph S. Alper, Paul R. Billings, Carol I. Barash, Jonathan Beckwith, and Marvin R. Natowicz. "Individual, Family, and Societal Dimensions of Genetic Discrimination: A Case Study Analysis." *Science and Engineering Ethics* 2, 71–88, 1996.

Kass, Leon R. *Toward a More Natural Science: Biology and Human Affairs.* New York: The Free Press, 1985.

Kitcher, Philip. *The Lives to Come: The Genetic Revolution and Human Possibilities.* New York: Simon & Schuster, 1996.

Lapham, E. Virginia, Chahira Kozma, and Joan O. Weiss. "Genetic Discrimination: Perspectives of Consumers." *Science* 274, 621–4, 1996.

Lipshutz, R. J., D. Morris, M. Chee, E. Hubbell, M. J. Kozal, N. Shah, N. Shen, R. Yang, and S. P. A. Fodor. "Using Oligonucleotide Probe Arrays to Access Genetic Diversity." *BioTechniques* 19, 442–7, 1995.

Lyon, Jeff, and Peter Gorner. *Altered Fates: Gene Therapy and the Retooling of Human Life.* New York: W. W. Norton, 1995.

Wilmut, I., A. E. Schnieke, J. McWhir, A. J. Kind, and K. H. S. Campbell. "Viable Offspring Derived from Fetal and Adult Mammalian Cells." *Nature* 385, 810–3, 1997.

ACKNOWLEDGMENTS

I am fortunate to have worked at the National Institutes of Health with many talented research fellows and students, including Craig Fisher, Stella Hu, Nan Hu, Lin Li, Mark Nelson, Angela Pattatucci, Chavis Patterson, and Sue Sabol. I also benefited greatly from collaborating with Jon Benjamin, Ben Greenberg, Frank Lucas, and Dennis Murphy at the National Institute of Mental Health, and from the statistical expertise provided by Stacey Cherny and David Fulker at the University of Colorado and by Leonid Kruglyak at the Whitehead Genome Center. Special thanks for information, encouragement, advice, and criticism are due to Michael Bailey, Dietmar Bengel, Ray Blanchard, Deborah Blum, Robert Cloninger, Paul Costa, Milton Diamond, Pablo Gejman, Elliot Gershon, Brian Gladue, Irving Gottesman, Jeff Hall, René Hen, John Hunsaker, Pamela Jacobs, Martin Kafka, Claude Klee, Peter Lesch, Simon LeVay, Nick Martin, Robert McCrae, Mike Miller, Jeremy Nathans, Richard Pillard, Robert Plomin, Maxine Singer, Margo Wilson, Ken Zucker, and Marvin Zuckerman. Most of all, I thank the many people who have volunteered for our research studies for their time, cooperation, and interest in our work.

—D. H.

———

I would like to thank Dan Thomasson and Marvin West at Scripps Howard for their support. My wife, Maru, is the source of great inspiration and great genes, for which our children and I are so very grateful.

We would both like to thank David Walbaum for helping with the manuscript, Julie Castiglia for being a terrific agent, and Betsy Lerner for her inspired editing.

—P. C.

INDEX

ABOUT THE AUTHORS

DEAN HAMER, PH.D., is Chief of Gene Structure and Regulation in the Laboratory of Biochemistry at the National Cancer Institute in Bethesda, Maryland. He is a graduate of Harvard University and a pioneer in molecular genetics who has published over 100 research papers. Award-winning writer Peter Copeland is managing editor of Scripps Howard News Service in Washington, D.C. This is his fourth book. Hamer and Copeland also wrote *The Science of Desire,* a *New York Times* Notable Book in 1994.